NASA
SATURN V

1967–1973 (Apollo 4 to Apollo 17 & Skylab)

COVER IMAGE: The three Saturn V stages.
(Kevin Woods)

© W. David Woods 2016

All rights reserved. No part of this publication may be reproduced or stored in a retrieval system or transmitted, in any form or by any means, electronic, mechanical, photocopying, recording or otherwise, without prior permission in writing from the Publisher.

First published in August 2016
Reprinted January, May and December 2017
Reprinted May and October 2018
Reprinted April and October 2019
Reprinted May and November 2020
Reprinted October 2022 and October 2024

A catalogue record for this book is available from the British Library.

ISBN 978 0 85733 828 0

Library of Congress control no. 2016930194

Published by Haynes Group Limited,
Sparkford, Yeovil,
Somerset BA22 7JJ, UK.
Tel: 01963 440635
Int. tel: +44 1963 440635
Website: www.haynes.com

Haynes North America Inc.,
2801 Townsgate Road, Suite 340
Thousand Oaks, CA 91361

Printed in India.

NASA
SATURN V

1967–1973 (Apollo 4 to Apollo 17 & Skylab)

Owners' Workshop Manual

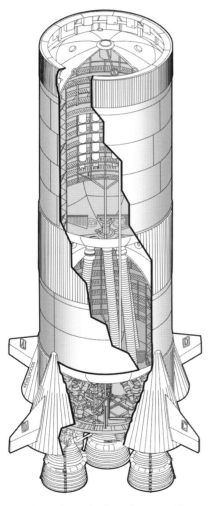
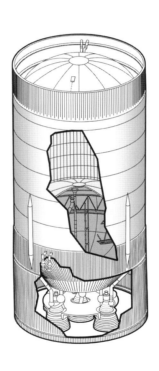
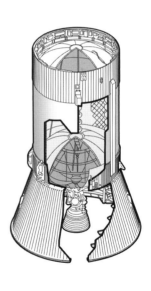

An insight into the history, development and technology
of the rocket that launched man to the Moon

W. David Woods

Contents

9	**Introduction**

10	**War-time to peace-time**	
A-4		13
Transplanting to the USA		15
Redstone and Jupiter		16
International Geophysical Year		17
Jupiter C – Juno I		18
Saturn I		18
NASA		19
Shifting plans		20
The Moon challenge		21
The vehicle defined		23
Down to work		24
All-up testing		25

28	**F-1: Brutal power**	
General description		31
Starting the F-1		43
Shutting down the F-1		45
The F-1 in context		46
Fishing for the F-1		47

48	**S-IC: The monster stage**	
Thrust structure		51
Fuel tank		53
Intertank structure		55
LOX tank		55
Forward skirt		57
Pogo suppression		57
Prelaunch fuelling		58
LOX loading		58
Bubbling		59
Tank pressurisation		59
Level sensing		60
Range safety ordnance		60
Retrorockets		60

62	**J-2: Efficiency in an engine**	
Centaur		65
Centaur to Saturn		65
Towards the J-2		68
J-2 development		68
Engine description		70
Starting the J-2		81
Engine shutdown		83
J-2X		83

BELOW The Apollo 17 space vehicle rolls out from the Vehicle Assembly Building on 28 August 1972. *(NASA)*

LEFT Five mighty F-1 engines power a Saturn V off the pad.
(NASA/Stephen Slater)

116	Instrument unit

Structure	119
Electrical power	122
Environmental control	122
ST-124 guidance platform	124
Azimuth alignment	126
Digital computer	127
Emergency detection system	130

132	The ascent of the Saturn V

Alignment	135
Propellant loading	135
S-IC pressurisation	136
S-II and S-IVB pressurisation	136
Automatic sequence	136
F-1 ignition	136
Lift-off	138
Dumb or smart guidance	142
All-weather testing	142
Max Q	143
Riding the rocket	145
Staging	146
Skylab 1's interstage	147
The interesting flight of Apollo 6	148
S-II pogo and Apollo 13	150
Mixture ratio change	151
The S-IVB's first burn	151
Translunar injection	152
Bound for the Moon	153
Lunar impact	153
The historic Saturn V	155

84	S-II: The troubled stage

Destruction	87
Airframe	91
Tankage	91
Common bulkhead	93
Insulation	94
Level sensing	95
Propellant utilisation	96
Tank pressurisation	97
Propellant dispersion	98
Ullage rockets	98
A singular stage	99

100	S-IVB: A stage to the Moon

Airframe	105
Tankage	107
Insulation	108
Level sensing and propellant utilisation	108
Engine prechill	109
Tank pressurisation and venting	110
Attitude control	112
Auxiliary propulsion system	113
Retro and ullage motors	114
Propellant dispersion	115
The S-IVB's legacy	115

156	Skylab 1

Dreams of a space station	158
NASA after Apollo	158
Air Force intentions	159
The orbital workshop	160
Birth of Skylab	161
Description	162
Skylab 1's near-disaster	166
The rescue of Skylab	167
Skylab's demise	172

Acronyms

AAP	Apollo applications programme
ABMA	Army Ballistic Missile Agency
AM	airlock module
APS	auxiliary propulsion system
ARPA	Advanced Research Projects Agency
ASI	augmented spark igniter
ATM	Apollo telescope mount
CDF	confined detonating fuse
CM	command module
CMG	control moment gyro
COI	contingency orbit insertion
CPU	central processing unit
CSM	command and service module
EDS	emergency detection system
EOR	Earth orbit rendezvous
G&N	guidance & navigation
GH_2	gaseous hydrogen
GOX	gaseous oxygen
GRR	guidance reference release
IGM	iterative guidance mode
IU	instrument unit
JSC	Johnson Space Center
KSC	Kennedy Space Center
LH_2	liquid hydrogen
LOR	lunar orbit rendezvous
LOX	liquid oxygen
LVDC	launch vehicle digital computer
Max Q	maximum dynamic pressure
MDA	multiple docking adapter
MOL	manned orbiting laboratory
MORL	manned orbiting research laboratory
MR	mixture ratio
MSC	Manned Spacecraft Center
MSFC	Marshall Space Flight Center
NAA	North American Aviation
NACA	National Advisory Committee for Aeronautics
NASA	National Aeronautics and Space Administration
OWS	orbital workshop
PU	propellant utilisation
RAM	random access memory
RP-1	rocket propellant 1
SCE	signal conditioning equipment
SLS	Space Launch System
SM	service module
SSME	Space Shuttle Main Engine
STDV	start tank discharge valve
TD&E	transposition, docking and extraction
TLI	translunar injection
USAF	US Air Force
VAB	Vehicle Assembly Building

Units

Force
g	acceleration load
lb-f	pounds-force
N	newton
kN	kilonewton
MN	meganewton

Time
hr	hour
min	minute
sec	second

Distance
ft	foot
m	metre
cm	centimetre
mm	millimetre
km	kilometre
nmi	nautical mile

Speed
ft/sec	feet per second
m/sec	metres per second
km/hr	kilometres per hour
km/sec	kilometres per second
rpm	revolutions per minute

Electricity
V	volt
W	watt

Power
kW	kilowatt
MW	megawatt
GW	gigawatt
bhp	brake-horsepower

Radio
MHz	megahertz

Angles
°	degrees
arc-min	minute of arc

Temperature
°C	degree Celsius
K	Kelvin

Pressure
psi	pounds per square inch
kPa	kilopascal
MPa	megapascal

Mass
lb	pound
kg	kilogram

Conversion factors

Distance
feet	0.3048	metres
metres	3.281	feet
kilometres	0.6214	statute miles
statute miles	1.609	kilometres
kilometres	0.54	nautical miles
nautical miles	1.852	kilometres
nautical miles	1.1508	statute miles
statute miles	0.86898	nautical miles

Velocity
feet/sec	0.3048	metres/sec
metres/sec	3.281	feet/sec
kilometres/hr	0.6214	statute mph
statute mph	1.609	kilometres/hr

Volume
US gallons	3.785	litres
litres	0.2642	US gallons
imperial gallons	4.546	litres
litres	0.22	imperial gallons
cubic feet	0.02832	cubic metres
cubic metres	35.315	cubic feet

Mass
pounds	0.4536	kilograms
kilograms	2.205	pounds
metric ton (1,000kg)	1.102	short ton (2,000lb)
short ton (2,000lb)	0.9072	metric ton (1,000kg)

Pressure
pounds/sq inch	6.895	kilopascals
kilopascals	0.145	pounds/sq inch

Force
pounds-force	4.4482	newtons
newtons	0.2248	pounds-force

Power
watts	0.00134	horsepower
horsepower	745.7	watts

Specific impulse
second	9.81	metres per second
metres per second	0.102	second

Acknowledgements

Like all my works on Apollo, this book stems from a personal fascination with the programme. However, it could not have reached this point without the help of a few good folk to whom I must express my thanks. Primary among them is Mike Jetzer, curator of the heroicrelics.org website. Mike's knowledge of the Saturn V and especially the F-1 engine is extraordinary and I am indebted to him for his archive, his photography and his exhaustive attention to detail and fact-checking. There is much in this book that he can take credit for.

Additional help with images came from Lee Hutchinson, Alan Lawrie, Ed Hengeveld, Scott Schneeweis, Phil Broad and Justin LaFountain. Stills were acquired from film footage supplied by Stephen Slater. JPL engineer Marc Rayman helped tighten up my explanation of specific impulse. I'd also like to thank the good folk of the Project Apollo Yahoo group who responded to a call for a document; in particular, Linden Sims and Bob Andrepont. Bob must also take credit for collating the huge number of PDF documents that have been a constant wellspring of information for me. Thanks to IBM Archives, Bezos Expeditions, the US Space and Rocket Center and NASA for images and archive material.

A book like this can only come about when a publisher puts faith in an author and I owe a huge debt of gratitude to Haynes Publishing and my commissioning editor Steve Rendle. Steve is always a gentleman and is a pleasure to work with.

Only on a very few occasions did I bypass the sage advice of my esteemed editor, David M. Harland. Except for those occasional lapses on my part, the quality of the text owes much to his skills and to his knowledge of space flight and the Apollo programme. Any errors, factual or grammatical, are therefore my own.

For me, writing a book requires the tolerance of my family. My sons, Stephen and Kevin, are always a delight to me and I am especially pleased that Kevin was asked to produce the illustrations for the book's cover. He now knows more about the Saturn V than he ever wanted.

Finally, for cups of tea and coffee, for encouragement, for tolerating my passion for space and for sharing this ride on spaceship Earth with me, I would like to thank my amazing partner, wife and best friend, Anne. It is to her, with all my love, that I dedicate this book.

Introduction

1967 was a tragic year for space flight on both sides of what was at the time a geopolitical battle for the high ground of space. January had seen the deaths of three American astronauts during a test of their oxygen-filled spacecraft. In April, a Soviet cosmonaut died when his re-entry craft's parachute failed to deploy at the end of a difficult inaugural mission. But on 9 November, the media gathered some 5km from Launch Complex 39 at the Kennedy Space Center to broadcast the first launch of the United States' newest and largest rocket, the Saturn V. Though they had already witnessed many missions being launched from Florida's space coast, they were not prepared for what was about to hit them.

As soon as the Public Affairs Officer passed 'nine' in his countdown to the moment of lift-off, he announced, "Ignition sequence starts." At the base of the 110m tall rocket, five giant F-1 engines came to life by spinning up their individual turbopumps. Each pump, rated at 53,000 horsepower, was about to draw kerosene and liquid oxygen from two giant tanks at a rate of over 2.5 metric tons per second and feed them into an adjoining combustion chamber. The pressure in the fuel line quickly rose and as it did so, it burst a cartridge to release a chemical that would react with the liquid oxygen and ensure the commencement of burning. With only one second remaining in the countdown, all five engines had settled into their task of producing nearly 60 gigawatts of energy and thereby sufficient total thrust to lift the 3,000-metric-ton vehicle into the sky.

As the Saturn V began to rise, the tremendous pressure waves caused by the F-1s raced across the Florida scrub, taking about fifteen seconds to reach the press corps and begin their assault on the newsmen.

"The building's shaking here," yelled Walter Cronkite in the CBS cabin. "Our building's sha… The roar's terrific." He and his team feared for the flimsy shed that surrounded them. "This big glass window is shaking and we're holding it with our hands. Look at that rocket go!" he hollered above the racket. "Part of our roof has come in here."

NBC were having an equally animated time. "We don't know if you can hear, ladies and gentlemen … but the NBC observation booth is literally being shaken apart." The

OPPOSITE The Apollo 8 space vehicle, AS-503, seen at dusk as the mobile service structure on the left is about to be pulled back. *(NASA)*

ABOVE The thunderous lift-off of AS-501, the first Saturn V launch and the start of the Apollo 4 mission. *(NASA)*

acoustic energy from the Saturn V wasn't just heard, it was felt. "Our tape recorder has been thrown to the floor by the roar of this mighty rocket as it continues to climb into the sky on its seven and a half million pounds of thrust. It's a beautiful sight – an unbelievable sight."

Apollo 4 took only 12 minutes to recede from view and enter Earth orbit. After the newsmen had regained their composure, the science editor for ABC News, Jules Bergman, reflected on the launch for his anchor, Frank Reynolds. "Frank, there's just never been anything like it. … not until you've felt your flesh vibrate … and your body thunder with the vibration from that rocket going could you sense the excitement. … And it worked better the first time than anything we've ever used."

And that became the legacy of the Saturn V rocket. Not only did it repeatedly take humans to the Moon; it carried out its task reliably and with spellbinding and impressive style. Its visceral roar could be heard across the state of Florida every time it launched; an unforgettable experience for the millions who witnessed it. This book will explain how it worked.

Chapter One

War-time to peace-time

The story of the Saturn V has its roots in the first half of the twentieth century when interest in rocketry blossomed within the advanced nations, especially the US, UK, Russia and Germany. It was fed by a rise in scientific knowledge and an accelerated technological capability. While military thinkers eyed the rocket as a means to project physical power, others saw it as the key to the dream of space travel. These opposing goals, destruction and exploration, would forever create an edgy tension within the world of rocketry.

OPPOSITE Apollo 13 sits poised at dawn on Pad 39A at the Kennedy Space Center in March 1970. On the left, the mobile service structure is a few metres away from its normal station where it would provide access to the vehicle. *(NASA-KSC)*

RIGHT Konstantin Tsiolkovsky made the first rigorous study of rocketry and space flight. *(Archive of the Russian Academy of Sciences)*

A handful of individuals and societies laid the groundwork of theoretical papers and practical studies on the application of the rocket as a means of achieving space travel. In 1861, William Leitch, a Scot, wrote an essay in which he correctly noted that the rocket principle of achieving motion in one direction by ejecting mass in the other, would work in a vacuum. A generation later, the first rigorous study of rocketry in space was carried out in Russia by Konstantin Tsiolkovsky, though his country gave little serious thought to the concept beyond small-scale military use.

Similarly in the United States, most rocket research was concerned with the use of solid propellants for military purposes. However, Robert Goddard, an engineering genius, worked with his team in seclusion to perfect technologies required for the liquid-fuelled rocket, leading to the launch of their first example on 16 March 1926. This was probably the first ancestor of the Saturn V. But Goddard was suspicious of outsiders and highly protective of his patents. He also suffered at the hands of the US media who, ignorant of the laws of physics, failed to understand the possibilities of his work and mocked him. US rocketry had to await another spark to ignite its development.

In Germany, Hermann Oberth had become a catalyst in the development of rocketry by promoting the concept of space flight, both publicly and within institutions. Between the wars, he became president of a rocketry club known as *Verein für Raumschiffahrt* (Society for Spaceship Travel) or *VfR* for short. Though it was an amateur society, the VfR attracted young engineers who could advance the state of the art; the most notable being Wernher von Braun whose skill and ability was also at the level of genius.

BELOW Robert H. Goddard (second from right with colleagues) brought practical experimentation to liquid-fuelled rocketry. This rocket flew on 19 April 1932. *(NASA-Goddard)*

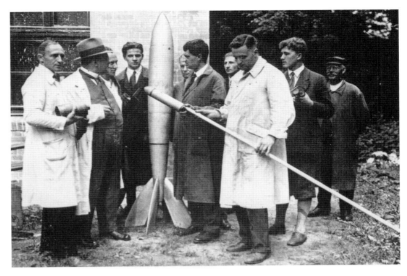

RIGHT **Members of the German rocketry club, the *Verein für Raumschiffahrt* (Society for Spaceship Travel). Hermann Oberth is to the right of the large rocket and Wernher von Braun is second from the right.** *(NASA)*

At this time, the German military were hamstrung by the Treaty of Versailles which concluded the Great War and placed severe restrictions on the type of artillery they could acquire. As a possible workaround and as an alternative to guns, they examined the use of rockets to deliver warheads over long ranges. For engineering expertise, they turned to the VfR and in 1936, with funding from a rapidly expanding war machine, a small engineering team headed by von Braun set up a military research station at Peenemünde on the island on Usedom, off the Baltic coast of Pomerania. Very soon, this fledgling facility was host to thousands of workers.

At Peenemünde von Braun developed a lineage of rockets that would lead directly to the Saturn V. The most important of these designs was known to the engineers as Aggregate-4 (their fourth major design). It was renamed by the Nazis as the V-2 for Vergeltungswaffe-2, variously translated as Retaliatory- or Reprisal Weapon-2 and it could send nearly a metric ton of high explosive to a target over 300km away.

A-4

The slender outline of the A-4, with its bullet-shaped body and sleek fins at the rear, became the archetype for the rocket in the popular imagination; an image that has endured despite decades of advancement in the field. But the working parts of the A-4 embodied many of the basic elements of liquid-fuelled rocketry that would eventually be expressed in the Saturn V.

The A-4 was a pointed cylinder 14m long and 1.65m wide at its greatest girth. A set of four rear-mounted fins with a wingspan of 3.5m

RIGHT **Diagram of the A-4 (V-2) rocket.** *(NASA/Mike Jetzer/Woods)*

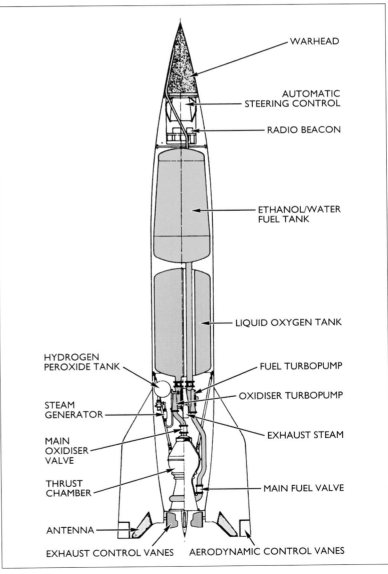

LEFT A-4 (V-2) being prepared for launch at Peenemünde, believed to be March 1942. *(German Federal Archive)*

provided aerodynamic stability. It comprised four distinct sections.

A nosecone formed the top 2m and carried the payload, usually a warhead. The control equipment occupied the next 1.4m. The main body was 6.2m long and contained a tank of ethanol/water fuel mounted above a tank of liquid oxygen. The 4.4m finned rear section contained the vehicle's motor, a device that profoundly influenced the future of rocketry because it used a turbopump to force the propellants into a combustion chamber. This

BELOW German drawing of an A-4 combustion chamber. *(Archives Dept., University of Alabama in Huntsville/Mike Jetzer)*

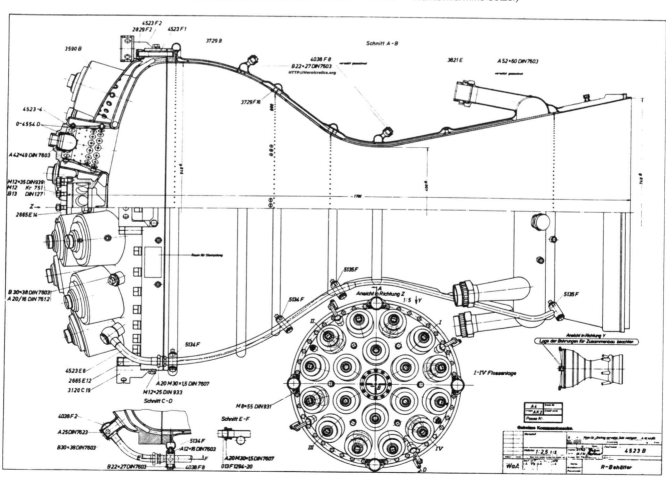

ABOVE The German rocket team at Fort Bliss in March 1946. Wernher von Braun is right of centre with his hand in his pocket. *(NASA/Mike Jetzer)*

turbopump was powered by having hydrogen peroxide, fed from a small storage tank, decompose into steam using a permanganate catalyst. The resultant high-pressure steam spun a turbine whose shaft then drove pumps for the fuel and oxidiser.

On its way to the combustion chamber, the ethanol/water fuel performed the additional task of cooling the combustion chamber and nozzle. It was fed to a manifold near the end of the engine, where it entered the double-skinned wall of the assembly. As it flowed upwards within the wall of the chamber, it was pre-warmed while it simultaneously cooled and protected the chamber walls from the 2,500 to 2,700°C heat of combustion. Fuel was also allowed to leak onto the internal surface of the chamber to further protect it by forming a boundary layer, a technique known as film cooling.

Steering was normally achieved by reference to gyroscopes in the control section. These generated steering commands that operated on a set of eight vanes at the bottom of the rocket. At the trailing end of each of the four fins was a small vane which provided aerodynamic steering. Four graphite vanes were mounted in the exhaust stream to deflect it for thrust vector control. The control section also housed an accelerometer that measured the rocket's changes in velocity. It sent a cut-off signal to the motor when a desired velocity was attained.

In the early days of the A-4, some people at Peenemünde viewed it as the first step towards space flight. The Nazi regime had no such visions, and instead devoted huge and scarce resources to the mass production of their V-2 terror weapon. Starting in September 1944, over 3,000 were launched with mixed military success. Probably the greater cost of the V-2 was paid by those who built them. At the Mittelbau-Dora concentration camp, slave labour was brutally engaged to manufacture the V-2 under dreadful, inhumane conditions. It is estimated that 20,000 workers died horrifically while being forced to build Hitler's miracle weapon.

Transplanting to the USA

As the Second World War came to an end, victorious power blocs scoured Europe for the technological spoils of war. Simultaneously, many Peenemünde engineers realised that their knowledge and skills would be of great interest to the Allied powers. A large group, led by Wernher von Braun, decided that the United States was the best option and so they arranged to hand themselves over to the American military rather than risk being taken

by the Soviets. Over the succeeding few years, Operation Paperclip transplanted 127 selected members of von Braun's team, along with their families, to the USA.

Initially, the German rocket team was installed at Fort Bliss in Texas while it helped the US Army with tests of captured A-4/V-2 rockets at White Sands in New Mexico. In 1949, the group was relocated to Huntsville in Alabama where two large Army arsenals, Huntsville and Redstone, offered much scope for growth. It was becoming clear that powerful rockets derived from the A-4 could carry nuclear weapons over very long distances and the US Army was keen to capitalise on the knowledge garnered by the Germans in this field.

Redstone and Jupiter

Under the direction of von Braun, the Huntsville team was tasked with the design and construction of a home-grown missile, the Redstone. This was essentially an improved version of the A-4. It had the same propellants in larger tanks and an upgraded motor supplied by Rocketdyne, a division of North American Aviation. Like the A-4, it was steered by vanes acting within the motor's exhaust along with vanes mounted on the aerodynamic fins. Chrysler was the prime contractor for this short-range (about 300km) surface-to-surface missile.

The Redstone first flew on 20 August 1953 and after a series of development flights it was declared operational in 1955. It was deployed as a tactical nuclear weapon carrier in West Germany until 1964.

Around this time, the development of US military rocketry began to diverge as inter-forces rivalry saw the US Air Force develop its own Thor and Atlas missiles. The basic technological lessons on rocketry had been learned and were being spread throughout the military industrial base. The Army and Navy meanwhile began a joint project to transform the Redstone into a

LEFT Redstone missiles were tested at Cape Canaveral, Florida. This is missile No. 1002 being prepared for launch on 16 May 1958. *(NASA-MSFC)*

medium-range (about 2,400km) missile. They called this Jupiter and it first flew on 1 March 1957. Jupiter missiles were then deployed across Europe and Turkey until 1963.

International Geophysical Year

As the International Geophysical Year (IGY, 1957–58) approached, all three US forces proposed launching a satellite as part of the nation's contribution to that scientific endeavour. However, to minimise the perception of a military-led effort, a plan from the Naval Research Laboratory (NRL) was selected. It would re-engineer a sounding rocket, the Viking, by adding stages that would turn it into a vehicle, the Vanguard, capable of placing a very small satellite into orbit.

Meanwhile, the USSR had been busy developing missiles based on German engineering skills that it had acquired at the end of the Second World War. By virtue of its nuclear weapons being substantially heavier than those of the US, it was forced to design missiles that were much more powerful than its US contemporaries. This became a distinct advantage when it too harboured plans to loft a satellite for the IGY. It had access to very capable rockets that could easily be turned towards space flight.

Because many in the West could not bring themselves to believe that the Soviets were capable of achieving space flight, they were shocked by the launch of Sputnik on 4 October 1957. Hence the USSR not only gained international prestige, it also caused the path of US rocketry to make a huge change of direction.

In reply to the Soviet success, the NRL attempted to get their Vanguard off the ground, the event being covered on TV for the world to see. Unfortunately, the launch was a spectacular failure. The US reeled in acute embarrassment as even their own press mocked their inability to match their rivals. However, von Braun had a solution ready and waiting.

LEFT Test launch of a Jupiter missile at Cape Canaveral on 20 October 1960. *(USAF)*

RIGHT In reply to the Soviet success with Sputnik, the US attempted to launch this Vanguard rocket on 6 December 1957. The attempt was an ignominious failure. *(NASA)*

ABOVE Launch of Explorer I by a modified Jupiter C vehicle, Juno I, on 31 January 1958. *(NASA)*

Jupiter C – Juno I

Wernher von Braun always had one eye on space flight. This was true whether he was designing the A-4 to enable the Nazis to hit London or the Redstone/Jupiter missiles to nuke Soviet forces across post-war Europe. It is therefore no surprise that it was he who formulated the US Army's proposal for the IGY satellite and he did his best to give it a non-military spin.

In support of the programme to build the Jupiter medium-range missile, a vehicle called Jupiter C (for Composite) was designed. Based on the Redstone this was a sounding rocket; that is, although it was intended to fly into space it was to pursue a suborbital trajectory, essentially up and down to perform research into the protection of warheads during re-entry. It used a new, more powerful fuel called Hydyne instead of the ethanol/water mix that had been inherited from the A-4, but the most important modification was the addition of two solid-fuelled stages to raise the payload's re-entry speed. However, on an appropriate flight path the three stages could take the vehicle's capability to a near-orbital speed of 7.8km/sec. By adding a fourth stage, it could accelerate a payload horizontally to achieve orbit. Again, to minimise the vehicle's military antecedence, the four-stage version was renamed Juno I.

It is interesting to note that a test of a Jupiter C on 20 September 1956 included an inert fourth stage which had been ballasted with sand. Had this stage been capable of firing, it would easily have put itself into orbit. But for the fact that the US Army was forbidden to arm this fourth stage in order to clear the way for the Vanguard project, the US would have launched a satellite more than a year before the Soviet success with Sputnik.

In the event, and in view of the failure of the NRL to launch Vanguard, the way was cleared for the Army, in association with scientist James van Allen and the engineers at CalTech's Jet Propulsion Laboratory, to use a Juno I vehicle to place the Explorer I satellite into orbit on 31 January 1958.

Saturn I

In 1956, the Huntsville team, by now known as the Army Ballistic Missile Agency (ABMA), had found itself left out of the effort to develop long-range missiles which had become the exclusive purview of the US Air Force. As an alternative, it looked towards using rockets to access space. Given the US government's expanding interest in the exploitation of space and their demands for accelerated booster development, it was thought that a large booster with a thrust of 6.7MN (1,500,000lb-f) would be useful.

Referred to as the 'Super Jupiter', it would use small engines that were already in their development cycle, arranged in a cluster to achieve the required thrust. The E-1 engine was originally chosen for a cluster of four because it was nearly ready for use, but then it was decided instead to cluster eight engines from the Jupiter missile in order to further accelerate development. This vehicle was named the

Juno V but some within Huntsville were already calling it Saturn.

Meanwhile, the Jupiter's engine was uprated and modified to make it more suitable for a clustered configuration. It was initially feared that having so many engines so close together would cause interference and overheating, but tests proved otherwise.

In order to quickly create the tankage to test the clustered rocket, engineers cobbled together tanks from the Jupiter and Redstone missiles rather than design and fabricate new tanks 6m in diameter. Although this arrangement produced an odd-looking vehicle, it became set in the blueprints as the first stage of the Saturn I and IB, the latter of which was used right through to 1975.

NASA

In October 1958, the US space effort was rationalised into a single civilian agency, the National Aeronautics and Space Administration, or NASA. Immediately this agency began to think about what it might do beyond getting a man into space with the Mercury programme. One committee, chaired by von Braun of the ABMA, looked at the vehicles available to give the US a lead over the Soviet Union in space capability. Its report included five categories of launch vehicle, the largest of which would cluster two or four F-1-class engines that were then in development and promised 6.7MN (1,500,000lb-f) of thrust. The committee believed such massive boosters could enable a large manned space station in less than five years and a base on the Moon in fifteen years.

Through 1959, ideas for a very large booster were studied, including what rocket should serve as the upper stage for the new eight-engined Juno V. One suggestion was to mount an entire Titan missile on top. Another proposed a new hydrogen-powered stage, the Centaur.

Meanwhile, the ABMA's large booster

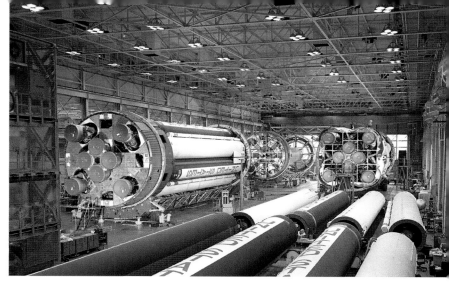

ABOVE A collection of S-I stages being constructed in Building 4705 in MSFC, taken on 13 January 1963. *(NASA-MSFC)*

BELOW Marshall Space Flight Center, the home of the Saturn V. *(NASA-MSFC/ Mike Jetzer)*

programme was renamed Saturn in February 1959, and in recognition of the fact that the ABMA was a unique resource of engineering talent, it was subsumed into NASA across the start of the new decade.

Furthermore, it was becoming clear that defence roles could be fulfilled using existing missiles like Atlas and Thor, while the Saturn programme was of more use to a rapidly expanding and ambitious NASA.

The transfer of the ABMA to NASA was set in stone by the creation of the Marshall Space Flight Center (MSFC or just 'Marshall') in Huntsville. On 1 July 1960, Wernher von Braun became its Director. Since its primary task was to develop a heavy-lift capability for civilian space goals, it became the birthplace of the Saturn V.

Shifting plans

Throughout 1960 and 1961, a snowstorm of designs and concepts for heavy-lift launch vehicles came out of MSFC as engineers reacted to shifting ideas about what the US was going to do in space. For the super-booster concept, the idea of a possible manned mission to the Moon led to paper schemes that called for a vehicle named Nova, an overwhelmingly huge rocket concept usually described as having eight to ten F-1s clustered in its first stage.

For Saturn, the odd-looking multi-tanked booster that had started life as the Juno V was now the Saturn I but there was much speculation about what would form its upper stages. A concept dubbed 'A' would use existing rockets for the upper stages but it was deemed too limiting. The 'B' concept envisaged an entirely new upper stage with conventional propellants (kerosene and liquid oxygen). This too was passed over. NASA's ambitions demanded more performance and the combustion of hydrogen seemed the way to achieve it. This was the 'C' option.

A rocket that burns liquid hydrogen can deliver much more energy per unit mass than any other common propellant. To make use of it, engineers were planning a range of Saturn upper stages dubbed S-II, S-III and S-IV, plus an S-V which was essentially the two-engined Centaur already in development. The S-IV and S-V would use the RL-10, a relatively small engine made by Pratt & Whitney. But the S-II and S-III would require a new hydrogen engine that would be an order of magnitude more powerful; the J-2.

With these building blocks, a range of configurations were laid out; the C-1, C-2 and C-3, each of increasing size and capability. But there was another option that would make very good use of the F-1 and J-2 engines. It rejected the S-I stage with its eight H-1 engines and replaced it with a much larger first stage that clustered initially four, then five F-1 engines. It wasn't as big as a Nova and it ought to have been ready by late 1967; far earlier than Nova. It could place a very useful 113 metric tons into low Earth orbit or send 41 metric tons to the Moon. Its paper designation was the Saturn C-5.

Through time, MSFC settled on three configurations of the Saturn vehicles to take forward: the C-1 would be used for early development flights, an upgraded C-1, called the C-1B would use the third stage intended for the C-5 and would therefore be able to help to test that component. Finally, there was the C-5 itself.

BELOW An early NASA infographic comparing the C-1, C-5 and Nova launch vehicles. *(NASA-MSFC)*

RIGHT SA-9 launches on 16 February 1965. The vehicle consisted of an S-I booster and S-IV second stage which lofted a Pegasus satellite and a boilerplate Apollo spacecraft with a launch escape tower. Once in orbit, Pegasus deployed a pair of 'wings' which operated as a huge micrometeoroid detector. *(NASA)*

The Moon challenge

By early 1961, teams within NASA were thinking very seriously about what it would take to get a human onto the Moon. It was one of the possible missions that their studies had constantly referred to, along with circumlunar flights and a permanent manned station in Earth orbit. They even had a spacecraft in mind for it called Apollo. Thus when President J.F. Kennedy needed a goal to display his nation's mettle to the world, some of the groundwork for a lunar landing had already been laid.

Then, as soon as the challenge of 25 May 1961 had been made, to land a man on the Moon and return him safely to Earth within the decade, the various NASA centres began to discuss ways of carrying out the mission. There were three scenarios in this often heated debate. Two required the Saturn C-5 while the simple brute-force method needed the Nova.

It took nearly 18 months to settle this issue, and in the course of the controversy the Nova option was dropped. Soon after, on 7 February 1963, NASA renamed the new advanced boosters for the Apollo programme. Those based on the Juno V design became the *Saturn I* and the *Saturn IB*. The name *Saturn V* was coined for the large C-5 super-booster design.

When it came to the 'mission mode' question, NASA rejected *Earth orbit rendezvous* (EOR), a technique that would have required the launch of two Saturn Vs for each mission. This would then have entailed a rendezvous in Earth orbit to form a large craft capable of both landing on the Moon and returning to Earth. It was EOR that defined the requirements of Launch Complex 39 at Cape Canaveral in Florida to handle a large throughput of launch vehicles as well as near-simultaneous launches.

LEFT The launch of AS-202 on 25 August 1966. This Saturn IB vehicle comprised an S-IB first stage, an S-IVB second stage and an Apollo spacecraft. Its purpose was to confirm that the system would be able to carry humans on its next mission. *(NASA)*

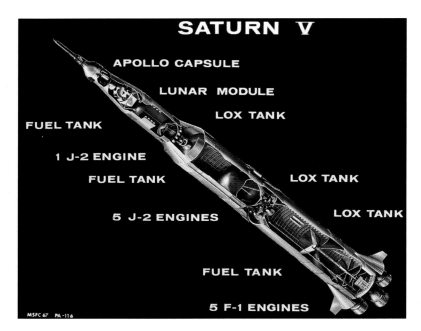

ABOVE Having settled down in its configuration of S-IC, S-II and S-IVB, the C-5 vehicle became the Saturn V, as shown in this NASA infographic. *(NASA-MSFC)*

BELOW The mission plan for the LOR mode required that a rendezvous be carried out in the Moon's vicinity. The weight savings that arose from LOR allowed a lunar landing to be achieved using only one Saturn V. *(Woods)*

In the event, NASA realised that landing a large and unwieldy craft on the Moon would be less than ideal and ironing out the bugs of EOR would likely delay the programme beyond Kennedy's deadline.

The chosen mission mode for the Moon flights was *lunar orbit rendezvous* (LOR), a method that minimised the mass of the spacecraft at every stage of the flight, discarding each piece once its role had been carried out. LOR required that a small, dedicated lander be specifically designed to touch down on the lunar surface while a mothership waited in lunar orbit. The entire Moon-bound craft wouldn't have to make the landing and only part of the lander would need to leave the Moon at the end of a visit.

The mass savings achieved by adopting LOR were enormous, to the extent that a lunar landing mission could be achieved using a single Saturn V. The engineers' decision in July 1962 to adopt this mode, and the acceptance by NASA HQ four months later, fired the starting gun in a race to procure, test and fly all the hardware necessary to achieve a lunar landing before 31 December 1969.

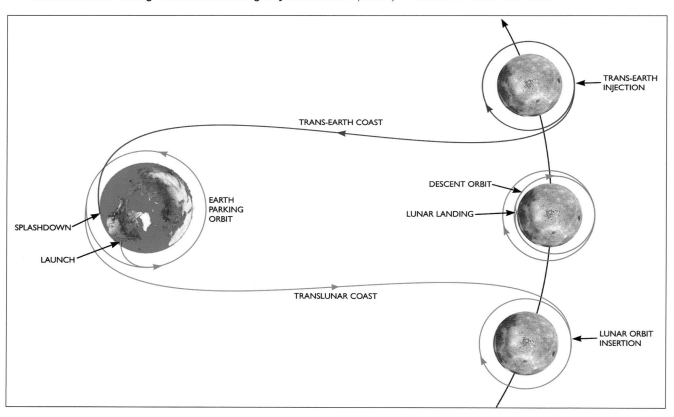

The vehicle defined

While NASA flailed around, working out how to get to the Moon, it nevertheless forged ahead, defining the important elements of the super booster. It had become increasingly clear that whichever mode was chosen, it would require not the Nova, but the C-5, later to become the Saturn V.

The first stage, the S-IC, would be huge at just over 42m tall. Most of its bulk would comprise a pair of tanks, each 10m in diameter, stacked one on top of the other with a cylindrical intertank to make room for their domed bulkheads. A rigid thrust structure would transmit the force of the five high-power F-1 engines to the skin of the vehicle. Although burning kerosene with liquid oxygen isn't the most efficient combination of propellants, it would provide the greatest raw power to get the whole vehicle and its payload into the upper atmosphere.

The second stage was originally envisaged as a four-engined stage for the C-3 variant. Initially intended to have the same diameter as the S-IB booster, the S-II second stage ballooned to become an advanced cryogenic rocket 10m wide to match the S-IC below it. It would also become a major headache both to NASA and its contractor. To minimise weight, its two tanks were closely joined via a common bulkhead which shortened its length to a little less than 25m. Power would come from five J-2 engines that burned the highly efficient combination of liquid hydrogen with liquid oxygen. It was the S-II that really made the Saturn V go, providing most of the velocity needed to get its payload into orbit.

ABOVE LEFT Apollo 8's first stage, S-IC-3, being hoisted within the Vehicle Assembly Building. It is missing its conical fairings, its fins and the nozzle extensions from its engines. *(NASA-KSC)*

ABOVE Saturn S-IC stages being assembled at the Michoud Assembly Facility in New Orleans, Louisiana. *(NASA-MSFC)*

LEFT The first flight-capable S-II stage being mated to the Apollo 4 S-IC and interstage. The stage is missing its aerodynamic fairings over the fuel ducts. *(NASA-KSC)*

RIGHT An S-II stage being assembled by North American Aviation at the Seal Beach plant in California. *(NASA-MSFC)*

FAR RIGHT Apollo 12's S-IVB in the vehicle assembly building having been mated to the rest of the vehicle. A temporary work platform has been installed within the forward skirt of the stage to allow access to its systems and to the instrument unit still to be attached. *(NASA-KSC)*

The third stage was already being developed for the Saturn IB vehicle and was called S-IVB, a name that revealed its origin among the plethora of paper designs that abounded as the Saturn was being defined. Its diameter was similar to that of the Saturn IB's booster at 6.6m. In its Saturn V configuration, its 18m length included a conical interstage that adapted its smaller diameter to the rest of the Saturn V. Like the S-II, a common bulkhead separated its two cryogenic tanks. Liquid hydrogen and liquid oxygen propellants were burned in a single J-2 engine. Its primary role on a Moon-bound Saturn V would be to accelerate its payload away from Earth.

Down to work

NASA worked with five major contractors to build the Saturn V; three dealt with a stage each, a fourth, Rocketdyne, built all the major engines, and IBM created the instrument unit which was essentially the brains of the rocket.

The responsibility for the S-IC went to Boeing, which gained its contract on 15 December 1961. The huge stages would be constructed in the Michoud Assembly Facility outside New Orleans, a manufacturing plant that had been used to build either wartime cargo ships or army tanks. It had already been brought out of mothballs for S-I stage construction. It had canal access to the Gulf of Mexico and on to Florida to allow the stages to be barged to NASA's new launch operations site near Cape Canaveral.

S-IC construction began in 1963 and MSFC shared the construction of the first few units with Boeing to prove manufacturing techniques under close NASA supervision. The first of these was S-IC-T, intended for proving the entire stage in one of the giant test stands at the Mississippi Test Facility.

The S-II contract went to North American Aviation on 11 September 1961, a decision that became controversial when the company subsequently acquired the contract for the Apollo spacecraft. There were worries about whether it could cope with the workload. A construction facility for the stages was provided by the US government at Seal Beach, California. The completed S-IIs were shipped through the Panama Canal to the testing and launching sites on the US East Coast. In the event, demands for weight reduction and various catastrophes that occurred in development made the S-II the pacing item for the Apollo programme.

The contract for the S-IVB, a stage intended for both Saturn IB and Saturn V, went to the Douglas Aircraft Company in December 1961. It was already producing the similar S-IV for the Saturn I. Douglas chose to build the stages at its own plant in Huntington, California. Testing was carried out at their facility near Sacramento and completed stages were shipped to the East Coast inside Super Guppy aircraft.

All-up testing

As NASA's contractors set about the daunting tasks that would stretch them in every sense, one of the agency's no-nonsense managers sent out a directive that ratcheted up the demands even further. There would be no incremental testing of the Saturn V, where engineers proved one system before adding another. Instead, the vehicle would be tested all at once with all three stages live and with a complete instrument unit controlling the whole assemblage.

Up to this time, both in the burgeoning rocket industry and the aircraft industry before it, engineers had pursued a piecemeal approach to the development of highly complex machines. A new aircraft would be carefully taxied along a runway and the pilot would subsequently be debriefed. Then the aircraft might be allowed to rise off the ground several metres, then more study and debrief, then a short flight before yet more analysis, eventually leading to where the full capability of the machine could be tested. Likewise in rocketry, especially multi-stage vehicles.

Through their wartime experience, the MSFC Germans knew how difficult it was to pre-empt all of the little things that could trip up a rocket crammed with parts that were invariably working on the edge of what materials could withstand. A succession of failed launches led to improvements that ultimately resulted in a reliable stage. It was then an appropriate time to place a live stage on top. This approach required the fabrication of multiple stages, just for testing.

But in late 1963, George Mueller, who had just taken up a job to direct NASA's manned space flight efforts, looked at the brutal end-of-decade deadline and the beginnings of a squeeze on the budget, and saw a disconnect between the two. He had just come from managing a part of the Air Force's missile programme where they were starting to employ all-up testing to reduce costs and speed up development.

Mueller correctly perceived that incremental testing of the Saturn V was a non-starter. The plan had been to launch an S-IC carrying

ABOVE S-IVB-505 and S-IVB-211 in the assembly and checkout tower at Douglas (later McDonnell-Douglas), Huntington Beach, California. *(NASA-MSFC)*

BELOW George Mueller (centre, with glasses) shares the joy of a successful launch for Apollo 11 along with senior NASA managers Charles W. Mathews, Wernher von Braun and Samuel C. Phillips. *(NASA-KSC)*

dummy second and third stages. But an S-IC stage cost so much and was so demanding to build, test and transport, that it could be argued that if the first example were to fly successfully, it would be a shame not to test a live second stage on top. Likewise with the third stage. With all-up testing, the entire vehicle could be proved to work with a single flight. This could save a lot of money and, more importantly in the light of the programme's deadline, a great deal of time.

Mueller's directive caused dismay, especially with the conservatively minded German engineers. There was so much in the Saturn V that was new and very powerful: the two most obvious examples being the sheer size and energy output of the S-IC and the novelty of hydrogen power in such a large booster as the S-II with its clustered engines. On the other hand, by the time of the first Saturn V test an S-IVB would have already flown as the second stage of a Saturn IB. However, the centre's director, von Braun, was muted in his criticism.

Mueller's logic was impeccable. Though all-up testing seemed risky, the alternative was doomed to fail because of time and cost constraints. Mueller argued that the all-up concept could work if preflight testing of the stages and their components was of a sufficiently high standard. In his scheme, two successful all-up flights of the Saturn V would be sufficient to prove that the vehicle was good and that astronauts should ride the third example.

Mueller's all-up plan turned out to be solid, and was one of the gutsiest decisions made during the Apollo programme. It is likely that Kennedy's deadline would never have been met had NASA not followed Mueller's lead. Instead, only four years later, AS-501 flew as Apollo 4 on a completely successful inaugural flight of the Saturn V. Every stage performed as intended and a huge wave of confidence washed over an organisation which, earlier in the year, had been wracked by tragedy.

The Saturn V is a magnificent demonstration of the ingenuity of humans and their ability to apply physical forces to an otherwise unattainable goal, in this case, one with a peaceful outcome. This book will describe all three stages of the rocket, the ground-breaking engines that powered it, and the instruments and computers that kept it under control.

THE BASICS OF ROCKETRY

In 1919, American rocket pioneer Robert H. Goddard had an article published which laid out his theories on rocket flight. He wrapped up with a short thought experiment in which an imaginary rocket might deliver a small flash powder charge to the Moon's dark hemisphere. Observation of the flash from Earth would provide a means of proving that the rocket had achieved the 'extreme altitude' of the lunar surface.

The newspapers picked this up and sensationalised the 'Moon rocket', ignoring the seriousness of the rest of his work. In January 1920, an editorial in the *New York Times* mocked Goddard: "That Professor Goddard … does not know the relation of action and reaction, and of the need to have something better than a vacuum against which to react – to say that would be absurd. Of course he only seems to lack the knowledge ladled out daily in high schools."

Unfortunately, the writer of this foolish editorial had excessively relied on his intuition and had assumed that in order for a rocket to move forward, its exhaust had to push against something. He was embarrassingly wrong and it took 49 years and the flight of Apollo 11 to prompt the *Times* to issue a correction.

Rockets do indeed rely on the relationship between action and reaction. Isaac Newton summed this up succinctly in his Third Law of Motion: For every action, there is an equal and opposite reaction. This means that if you were to lean against a wall, the wall would be equally pushing against you.

Imagine the scenario where you are in a rowing boat in the middle of a becalmed pond. Instead of oars, you've been bequeathed a pile of bricks. The problem is to propel yourself to the shore using the bricks. One solution is to lift a brick and throw it in a particular direction. As you do so, the boat moves in the opposite direction and were it not for the viscosity of the water slowing the boat to a stop, it would continue until it reached the shore. Each brick helps the boat to move a little further and if you can muster up the effort to throw the brick with greater speed, it will likewise impart more speed to the boat. At the moment of throwing the brick, you are applying a force to it. By Newton's law, the brick is applying a force to you and hence into the boat.

Space rockets are no different except that, instead of throwing out bricks with a speed of several metres per second, a rocket expels gas at a speed of a few kilometres per second. Chemical rockets usually achieve this through the process of combustion. If a substance is burned, whether a solid or a liquid, it will produce a gas which will attempt to expand and occupy a far greater volume. If we arrange for that gas to expand in one direction, it will impart a force in the opposite direction, even in a vacuum.

The general arrangement for a liquid-fuelled rocket is to bring a fuel and an oxidiser together in a chamber in which they can combust. If the chamber were closed, then the internal pressure would dramatically rise due to the gas attempting to expand and this pressure would be exerted equally on all walls. By having an opening in the chamber, the expanding gas is able to exit. For a moment at least, the pressure inside the chamber is unbalanced, and like an inflated balloon that is let go, the unbalanced force will try to move the chamber.

If we now arrange for the chamber to receive a constant flow of combustible chemicals, known as propellants, into one end of the chamber and allow exhaust gas to leave the other, then we have a means to achieve constant and controllable thrust for as long as the supply of propellant lasts.

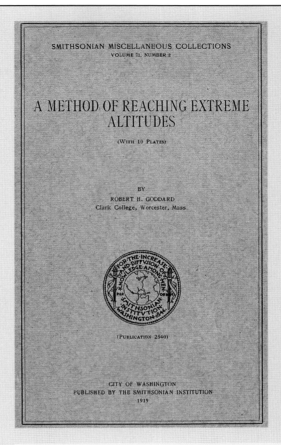

LEFT The cover of Robert Goddard's book that described his work and his thoughts on the possibilities of space flight.

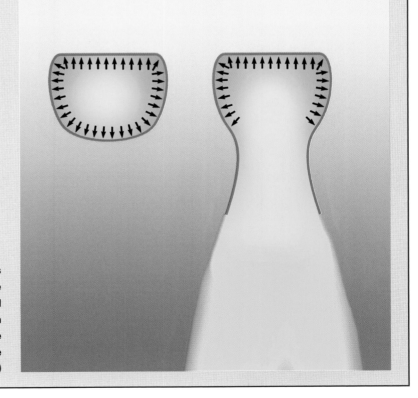

RIGHT Combustion in a closed space causes pressure to be applied equally around the chamber's walls. If the combustion is carried out in a chamber with one end open, then an unbalanced pressure will push against the other end, exerting an unbalanced force to the chamber. *(Woods)*

Chapter Two

F-1: Brutal power

No component of the Saturn V left a greater impression on those who witnessed it than the astonishing F-1 engine. At the time it was developed, a single example could generate as much thrust as the Saturn IB rocket, then the largest in the US which itself required the clustering of eight lesser engines. Harnessing such power taxed the engineers at Rocketdyne for over five years as they struggled to tame the capricious engine while, one after the other, early models displayed a remarkable tendency to consume themselves on the test stand.

OPPOSITE Nestled in the depths of a thrust chamber formed from hundreds of tubes is the copper-faced injector of the F-1 engine. Through 2,816 holes, over 2.5 metric tons of propellant was pumped each second, generating over 10 gigawatts of power. *(Lee Hutchinson)*

The F-1 began as a US Air Force project in 1955 when it became apparent that if space was to become a military high ground, then expected payloads like manned reconnaissance platforms and lunar bases would need some serious heft to get off Earth. Existing rocket engines were clearly too puny for the task.

The project's initial goal was to investigate the possibility of a 4.5MN (1,000,000lb-f) engine by stretching what was then known about how to build rocket engines. In the event, the USAF chose instead to rely on the rockets that had been developed as nuclear-tipped missiles; the Thor, Atlas and Titan vehicles. It had become clear that reconnaissance and electronic monitoring could be achieved more cost-effectively through the use of much lighter unmanned satellites.

In 1959, when it contracted Rocketdyne to create an enormous 6.7MN engine (1,500,000lb-f), NASA essentially inherited the Air Force project. The agency figured that by clustering this engine, it would be able to achieve boosters with astonishing lifting capability, eminently suited to its plans for manned space flight. Interestingly, it would be over two years before a vehicle was envisioned that would use this engine, the Saturn V.

By April 1961, a test of a prototype had achieved a peak thrust of 7.3MN. However, engineers were discovering the limitations of simply scaling up designs. Technical problems included exploding turbopumps and split fuel pipes in the thrust chamber. Combustion instability, already the bane of the engine designer, was also wreaking havoc with the development of the F-1.

Combustion instability was a product of the extremely dynamic environment in the combustion chamber of a liquid-fuelled rocket engine. Within the F-1's chamber, over 2.5 metric tons of propellant was being sprayed through the injector each and every second under very high pressure. After a short distance, it combusted and its temperature rose to over 3,000°C, creating extreme pressures in the chamber as the expanding exhaust gas quickly accelerated to speeds approaching 3,000m/sec. It didn't take much for minor pressure oscillations to quickly build up to ferocious levels. Then on 28 June 1962, it led to the complete destruction of an engine on the test stand.

The engineers had no means by which to directly study combustion instability and therefore no one had a good theoretical understanding of the problem. The normal empirical approach was to change something in the design of the injector and see if it helped. This had eventually worked for previous engines and was the only approach open to the F-1 engineers, despite being horrendously costly and inefficient.

What did help was a trick learned during the development of the H-1 engine for the Saturn I and IB. Rather than running engines for a long time and waiting for instability to occur, engineers forced the matter by implanting a bomb consisting of a small quantity of explosive (13.5 grains) encased in nylon. Mounted to the injector face but with its tip in the combustion zone, this bomb took about a second to lose its casing and ignite. When it did, the resulting pressure wave disturbed the conditions in the chamber sufficiently to induce combustion instability. The engineers then looked to see how long it took for the engine to naturally return to normal operation. Using this method along with their cut-and-try approach, they eventually reduced the instability to less than one tenth of a second.

This painstaking approach led to the propulsion engineers finally taming the F-1 in

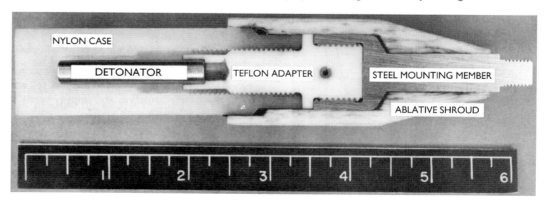

RIGHT A cutaway example of a combustion chamber bomb. *(NASA/Mike Jetzer)*

RIGHT An early F-1 engine being test fired at a stand at Edwards Field Laboratory in California. *(NASA-MSFC)*

the mid-1960s. The payoff was that across its operational lifetime from November 1967 to May 1973, 65 examples of this engine, the largest single-chambered liquid-fuelled rocket engine ever built, powered thirteen flights while turning in a near flawless performance.

General description

The F-1 engine burned a combination of RP-1 fuel with liquid oxygen (LOX) as an oxidiser to produce 6.7MN (1,500,000lb-f) of thrust in a single chamber. RP stands for

BELOW Diagram of the F-1 engine coloured to show the major fluid flows. The inset version shows how the shape of the high-pressure ducts changed to remove bellows and internal tie-rods from their construction, thereby simplifying it. The U-shape of the rigid duct accommodated any vibration between the turbopump and the thrust chamber. *(NASA/Woods)*

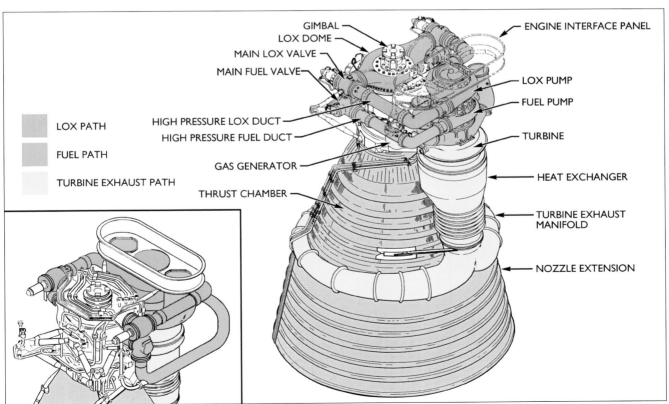

LEFT The F-1 thrust chamber was formed from a network of pipes or tubes through which fuel was passed prior to being consumed. Strengthening bands were brazed to the outside of the chamber to contain the pressures within.
(University of Alabama Archives/Mike Jetzer)

'rocket propellant', a type of kerosene which had been highly refined. This gave it a more predictable and consistent performance while also reducing the concentration of substances, particularly sulphur, that would have a deleterious effect on the engine through its test firings and final operation.

The engine was dominated by its thrust chamber, a beautifully sculpted piece of hardware where combustion took place at 3,300°C. The exhaust gas exited through the bell-shaped opening at up to 3km/sec. In the process, immense thrust was produced that acted in the opposite direction. The force of one engine was equivalent to the downward force of a 690-metric-ton block on Earth's surface.

Few metals can withstand the temperatures within an F-1 combustion chamber without some form of protection. The chosen solution, in common with many liquid-fuelled engines going back to the A-4, was to actively cool the interior. To this end, the chamber was constructed as a network of pipes (or *tubes* as Rocketdyne referred to them) through which 70% of the RP-1 fuel flowed for *regenerative cooling*.

The top of the chamber formed a cylinder slightly over a metre wide, capped by the injector, a steel plate faced with copper rings in which holes had been drilled in the manner of a showerhead. Each ring alternated between orifices for LOX and for fuel. Combustion occurred a few centimetres away from the face of the injector and the rapidly expanding exhaust then squeezed through a slight narrowing of the chamber before being allowed to expand in the bell-shaped nozzle whose cross-sectional area increased tenfold.

Above the injector was a dome that

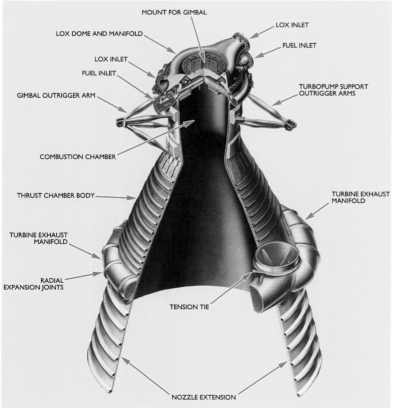

LEFT Cutaway drawing of the F-1 thrust chamber and nozzle extension with ancillary fitments.
(NASA-MSFC/Vince Wheelock/Mike Jetzer/Woods)

transmitted the force of the engine, via a gimbal bearing, to the rocket's structure. The gimbal bearing was a universal joint that allowed the engine to pivot so that the rocket could be steered. The dome also acted as a manifold to accept LOX and direct it to the LOX rings of the injector. Fuel entered the thrust chamber via another manifold that was wrapped around the edge of the injector.

During engine operation, the combustion of the propellants generated a pressure against the injector of 7.8MPa (1,125psi), about 65 times atmospheric pressure. It was this pressure that gave the engine the majority of its thrust. To maintain the flow of propellant into the chamber and stop it from being forced back into the supply system, it needed to be delivered at an even higher pressure. This was achieved by two powerful pumps that forced the liquids into the chamber at a rate of 1.8 metric tons of LOX and 0.8 metric tons of RP-1 every second (a mixture ratio of 2.27:1, LOX to fuel). These pumps were driven by a 53,000-brake-horsepower turbine, itself powered by burning propellants in a separate chamber called the gas generator.

ABOVE View into the F-1's thrust chamber towards the injector plate. At the point where the area of the expanding bell reaches a ratio of 3:1 compared to the throat, the number of pipes doubles from 178 to 356 as each one is bifurcated, or split in two. *(Mike Jetzer-heroicrelics.org)*

BELOW Cutaway drawing of the propellant delivery systems mounted at the top of the F-1's thrust chamber. *(NASA-MSFC/Vince Wheelock/Mike Jetzer/Woods)*

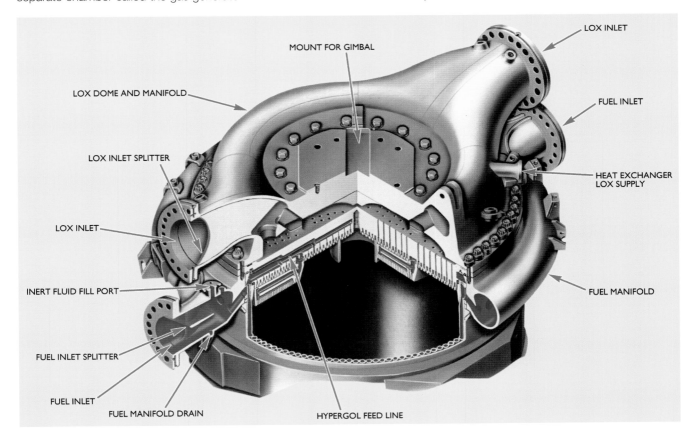

ABOVE Detail of the bifurcation of the thrust chamber pipework at the point where the bell's expansion ratio reaches 3:1. *(Mike Jetzer-heroicrelics.org)*

Once the exhaust from the gas generator had done its job of powering the turbopumps, it had two remaining tasks. The first was to pass through a heat exchanger where it warmed the gases that were used to pressurise the propellant tanks during flight. The second task for the turbine exhaust was to protect the large nozzle extension that was attached to the end of the thrust chamber's nozzle. This extension increased the expansion area of the whole nozzle from 10:1 to 16:1, thereby improving the efficiency of the engine, especially in the near vacuum of the upper reaches of the atmosphere.

Thrust chamber

Engineers chose to build the walls of the 2.9m wide × 3.35m high thrust chamber out of Inconel X750, a nickel alloy noted for its strength at high temperatures. This was in the form of pipework which allowed the chamber to be cooled by feeding high-pressure fuel through its passages prior to combustion.

Two inlets on opposite sides of the fuel manifold directed 70% of the RP-1 to 89 pipes that took it down towards the end of the nozzle. Another 89 pipes brought it back up. At the point along the nozzle where its area had expanded by a ratio of 3:1, each pipe was split in two, or *bifurcated*, to maintain its cross section. Therefore, from this point on, there were 178 *down* pipes and 178 *up* pipes. At the very end of the nozzle, where the nozzle extension was attached, a manifold directed fuel from the down pipes into the up pipework.

In order to seal any gaps between adjacent tubes, the entire network of piping was formed into a single unit by brazing; a process similar to soldering but at a much higher temperature whereby another metal is melted and allowed to flow between the two pieces to be joined without also melting them.

To achieve a consistent braze along all 900m of joints, the entire chamber was placed in a giant, shaped furnace, known as a *retort*, which performed the job in a single operation. This avoided localised heat distortion and permitted an oxygen-free atmosphere within the furnace whilst also allowing the metalwork of the piping to be age-hardened. Strengthening bands were hand-brazed to the outside of the pipework to contain the pressure of combustion during operation. Finally, the top of the chamber was capped by the injector.

Injector

The task of the injector was to introduce fuel and oxidiser into the combustion chamber and mix them as evenly as possible a short distance from its face. This would then promote their efficient combustion. It was formed from a large

BELOW A retort where F-1 thrust chambers were brazed. *(NASA/Rocketdyne)*

disk of steel, including a supporting flange, that had a 1m circular face made from 29 copper rings. Copper was used for the face because its high thermal conductivity meant that hotspots were less likely to form.

These rings alternated between the delivery of RP-1 fuel and LOX through 2,816 holes drilled at angles in such a way that the LOX jets impinged a few centimetres away from the face while the fuel jets impinged a short distance beyond. This was one of many schemes used for the patterns of injector impingement and was known as *like-on-like* or *biplanar* impingement.

A series of copper baffles, some concentric and some radial, protruded from the face of the injector and divided it into 13 compartments. Exhaustive tests had shown these to have a highly beneficial effect on the engine's combustion stability as they served to dampen the ferocious pressure waves that would otherwise build up within the chamber and destroy it. They were cooled by having fuel fed through internal channels which led out through small holes that fed into the chamber.

Each of the compartments between the baffles included two inlet ports that had been placed among the fuel orifices (with another one in the central compartment). These allowed a mixture of fuel and hypergolic fluid to be injected into the chamber in order to begin large-scale combustion during the engine's start sequence.

Gas generator

The F-1 needed an internal source of power for the propellant pumps. This came from the consumption of 2.9% of the available propellant which was burned in a small steel combustion chamber called the gas generator. The mixture ratio within this chamber was kept extremely fuel-rich (0.416:1, LOX to fuel) in order to keep temperatures within the generator low enough not to melt either its walls or the internal components of the turbine. With combustion taking place at a temperature of only 800°C, the

ABOVE Close up view of an F-1 injector and detail of the baffles that suppressed combustion instability. Fuel was passed through the baffles and exited into the chamber through the holes drilled along their ridges to cool them. The orifices that look like screw heads are where fluid from the hypergol cartridge exited to spontaneously react with LOX already present in the chamber, thereby ensuring ignition. *(Mike Jetzer-heroicrelics.org)*

BELOW Diagram of the layout of fuel and LOX rings in the F-1 injector. *(Woods)*

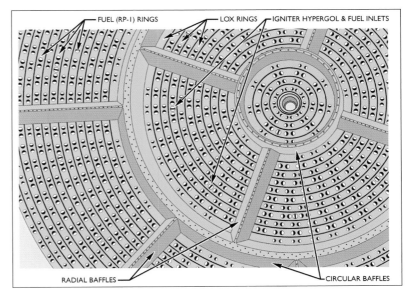

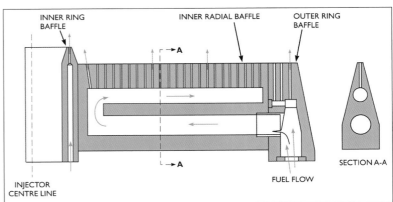

RIGHT Cross-section of an inner radial baffle to indicate the route taken by the fuel used to keep it cool. *(NASA/Woods)*

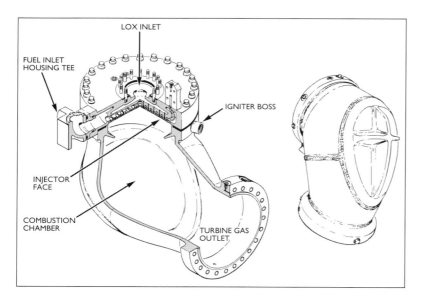

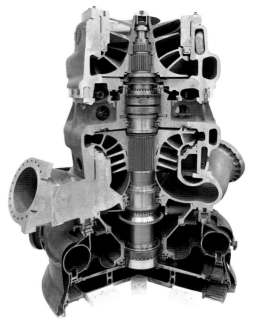

ABOVE Cutaway diagram of the gas generator. Fuel and LOX were burned in a fuel-rich mixture to create a gas to drive the turbine which powered the pumps. *(Rocketdyne/NASA)*

ABOVE RIGHT Cutaway of an F-1 turbopump. The turbine is at the bottom, fuel pump is in the middle and LOX pump is at the top. The turbine created over 53,000bhp (nearly 40MW) to turn the single shaft at 5,500rpm. *(Mike Jetzer-heroicrelics.org)*

BELOW Cutaway diagram of the F-1 turbopump. *(NASA/Woods)*

resultant exhaust gas contained much unburnt fuel. It was fed directly to the turbine section of the turbopump.

Turbopump

The turbopump was mounted alongside the cylindrical section of the thrust chamber with their axes parallel. It comprised three major elements: turbine, fuel pump and LOX pump in that order, bottom to top, all on a single shaft

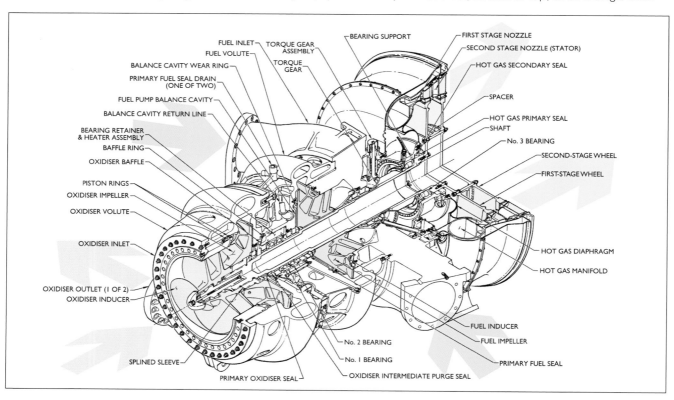

that ran from one end to the other. The order was chosen to separate the hottest section, the turbine, from the -183°C chill of the LOX pump. The turbine therefore directly drove the pumps without any gearing and with all the rotating parts directly affixed to the shaft.

The turbine section was a two-stage affair; i.e. it had two turbine wheels with vanes which were turned by the exhaust from the gas generator. Between the wheels, a stationary set of vanes prevented the gas from gaining rotary momentum. Each wheel extracted part of the energy from the gas flow to take the shaft to a rotational speed of 5,500rpm. Having given up much of its energy to the turbine, the gas departed downwards into a heat exchanger and on into the engine's nozzle extension via a large wrap-around exhaust manifold.

The fuel pump was fed from twin ducts that led from the fuel tank, with each duct containing a prevalve. To prevent cavitation (see box on page 38) it operated in two stages comprising an inducer and an impeller. The inducer partially raised the fuel's pressure and the impeller completed the task.

The impeller discharged the fuel at high pressure into a wraparound cover called a *volute* which had two outlets mounted on opposite sides. Having opposing outlets meant that any pressure change at the outlets was balanced and wouldn't impart a sideways or radial force on the shaft.

The fuel pump had its inlets below its outlets so that it was pushing fuel upwards, thereby placing a large downward force on the shaft.

The LOX pump was at the top of the turbopump assembly with a single large inlet at the end of the shaft. LOX entered via a large-diameter duct that ran from the LOX tank through the fuel tank and a prevalve directly into the inducer. Like the fuel pump, the LOX pump used an inducer followed by an impeller to stage the rise in pressure without permitting cavitation. High-pressure LOX was then discharged into a volute with two opposed outlets that minimised radial forces on the shaft.

Since LOX was entering the pump from the

ABOVE Close-up view of the vanes on one of the two turbine wheels in an F-1 turbopump. *(Mike Jetzer-heroicrelics.org)*

BELOW Two inducers from the F-1 turbopump. These are used in the first stages of each pump, raising the pressure of the liquids in readiness for the impellers as they turn clockwise. The LOX inducer is in the foreground with the fuel inducer behind. *(Mike Jetzer-heroicrelics.org)*

BELOW The LOX impeller from the F-1 turbopump. Spinning clockwise, LOX entered from the top and was driven down, to be pushed out centrifugally by the convex side of the lower vanes into the high-pressure volute. *(Mike Jetzer-heroicrelics.org)*

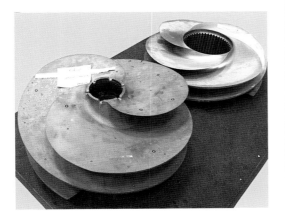

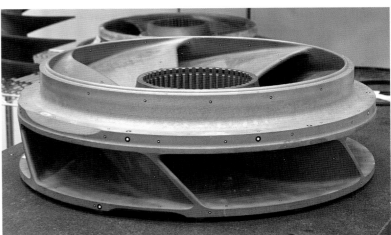

CAVITATION

If a boat propeller runs through water too fast, vapour bubbles or cavities can form in the liquid on the low-pressure side of the blades. This is an example of a phenomenon called *cavitation* and it can also occur on the low-pressure side of impeller blades within liquid pumps. Then, when the liquid pressure rises, the cavities collapse and the resulting shockwaves can damage the material of the impeller. Cavitation can therefore cause serious erosion of propellers and impellers in any kind of liquid. They are usually a sign that the blade is being forced to raise the pressure of the liquid too rapidly. Cavitation in a pump can be avoided by using multiple stages to raise the pressure in the fluid more gradually.

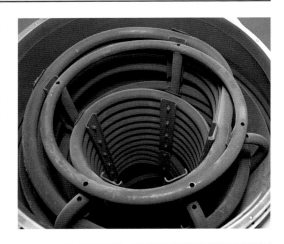

RIGHT Heat exchanger from the F-1 engine. Having passed through the turbopump, hot turbine gas passed over these coils, warming their contents, LOX and helium. *(Mike Jetzer-heroicrelics.org)*

RIGHT An F-1 engine mounted to an S-IC stage for testing. The lower section below the wraparound exhaust manifold was a bolted-on extension that improved the engine's efficiency. *(NASA-MSFC)*

top, the pump acted to push it downwards, exerting a large upward force on the shaft. This was in the opposite direction to the force exerted by the fuel pump and therefore the two forces were partially balanced, minimising the axial loads on the shaft's bearings. To further help with this balancing, a cavity within the high-pressure fuel volute allowed this pressure to act on the end face of the fuel impeller and exert an additional downward force.

Given that a single rotating shaft ran through the turbopump, a complex arrangement of bearings and seals was required to separate the three major sections while allowing free rotation. When the engine was running, the turbopump bearings were lubricated and cooled using the RP-1, a weight-saving measure that precluded a separate supply of of oil. On the ground prior to launch, ramjet fuel was pumped in for this purpose, since its properties were more suited to lubrication. Heaters were included for the bearings next to the LOX pump to prevent condensation and ice build-up prior to launch.

The technologies used for the seals varied according to their position within the system. Carbon seals were common, with springs pushing segments of carbon against the shaft. Leakage around the seals was allowed to drain away. Some seals in the LOX pump used Teflon and a similar substance called Kel-F which used chlorine as well as fluorine in its chemistry. These wouldn't react with the oxygen. So-called *labyrinth* seals used multiple intermeshing rings that made it difficult for fluids to pass merely by making the escape route tortuous.

Heat exchanger

Having powered the pumps, the turbine gas departed from the bottom of the turbopump unit into a heat exchanger where it passed over coils to which it gave up some of its heat. One set was used to warm helium gas from cold storage bottles in the LOX tank. This gas was then fed to the fuel tank to pressurise it. Another set of coils heated a feed of LOX to turn it into gaseous oxygen (GOX) to be fed into the top of the LOX tank in order to maintain its pressure as the liquid level dropped during flight.

The final task for the turbine exhaust was to protect the nozzle extension.

Nozzle extension

After the engine had been installed at the base of the S-IC stage, a single-piece extension was added to the end of the thrust chamber's nozzle. This increased the expansion ratio of the engine from 10:1 to 16:1 in order to improve its overall efficiency.

When a rocket engine operates in a vacuum, its exhaust plume wants to expand in all directions on departing the nozzle. Any gas that leaves in a direction other than rearward will impart a less than ideal reaction force on the rocket. The function of the flared nozzle is to cause as much of the gas as possible to travel rearward, thereby increasing the efficiency of the engine. This was the purpose of the nozzle extension.

However, there are limitations to this arrangement, a major one being the physical space available to house a large nozzle. Another is that as the nozzle is made larger, it raises the overall weight of the engine. There will come a point where the gain in thrust from nozzle enlargement is cancelled by the additional weight.

Also, much of the flight of the F-1 was through Earth's atmosphere. At sea level, the atmospheric pressure tends to confine the expanding exhaust gas to a narrow stream – as is shown by the long flames that emanated from a rising Saturn V soon after launch. At that time, there is little benefit from a greatly extended nozzle. With all these factors in mind, a 16:1 nozzle extension was chosen as the most appropriate compromise of available space, weight and the fact that it operated in the atmosphere.

The extension consisted of a double-walled cone of nickel alloy with a flange at its narrow end to allow it to be bolted to the thrust chamber. Its inner wall was lined with overlapping steel shingles for thermal protection. It didn't benefit from the regenerative cooling used in the thrust chamber's nozzle, but instead was protected from the blast of the main exhaust by a film of much cooler gases that had powered the turbine, now further cooled by their passage through the heat exchanger.

Upon leaving the heat exchanger, the turbine exhaust gas entered a large wraparound manifold that fed it into the nozzle through a series of slots just above the extension. The gas was also fed into the space between

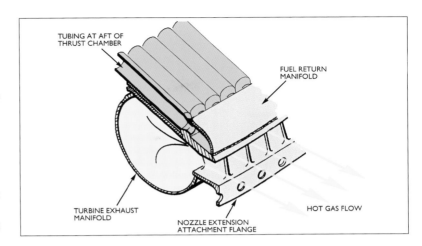

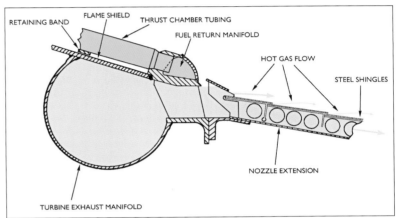

TOP Drawing of the periphery of the F-1's thrust chamber without the nozzle extension attached. This shows how turbine exhaust gas left the wraparound manifold. *(NASA/Woods)*

ABOVE Drawing to show how the turbine exhaust gas leaves the nozzle extension to form a protective boundary. *(NASA/Woods)*

BELOW Close-up view of the junction between the thrust chamber and the nozzle extension to show the slots where the turbine exhaust gas exits the manifold and enters the nozzle. *(Mike Jetzer-heroicrelics.org)*

LEFT An F-1 engine being tested at Edwards Field Laboratory in California. The dark, sooty turbine exhaust that is protecting the nozzle extension is clearly visible prior to it being consumed by the far hotter gases within. *(NASA/Mike Jetzer)*

the extension's double wall, from where it could penetrate between the shingles. This arrangement formed a relatively cool protective boundary layer across the inside of the extension. It can be seen in footage of the Saturn V's launch where a film of dark, sooty gas is seen to leave the exit plane of the nozzle. After a short distance, the fuel-rich turbine exhaust is consumed by the far hotter main exhaust gas. Of course, by that time, it had done its job.

Main control valves

One of the major advantages of a liquid-fuelled rocket engine is that it is straightforward to shut it down merely by turning off the flow of propellant through the use of valves in the feed lines, much like a water tap. Each propellant pump in an F-1 engine had two diametrically opposed outlets that fed high-pressure propellant to a pair of ducts; four in all, two for fuel and two for LOX. All four led to their respective manifolds at the injector. The main valves were immediately prior to the manifold inlet; two for fuel and two for LOX, one on either side. As well as controlled shutdown, an equally important task for these four valves was to control the start-up of the engine, only feeding propellant when appropriate in the ignition sequence.

The main LOX and fuel valves were of the poppet type. A disk, known as a *poppet*, was held against an aperture using spring force. A long rod connected the poppet to the actuating mechanism; in this case a hydraulically powered piston. There was a second sealed plug built into the poppet structure which balanced the internal forces on the valve. In other words, the pressure acting on the poppet was balanced

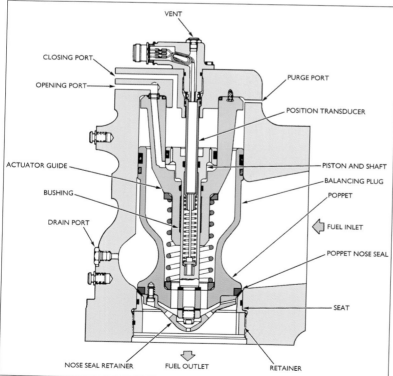

LEFT Cross-sectional diagram of the F-1 engine's main fuel valve. *(NASA/Woods)*

by the pressure acting on the plug. This made movement of the poppet easier.

The LOX valve had an extra function built into it. This was to sequence the operation of the valve with respect to another that supplied propellant to the gas generator. Essentially the gas generator wouldn't receive propellant, and therefore the turbopump couldn't get up to speed, until the LOX valve had opened to a pre-determined degree. This was achieved by a simple mechanism.

The valve's operating rod included an extension. Once the actuating piston had moved by 16.4% of its full travel, the tip of the extension rod pushed against a spring-loaded flap, opening it. This flap was a sequence valve that allowed hydraulic fluid through to open a supply of propellant to the gas generator.

Ancillary valves

Various valves around the engine worked with the main valves to control start-up, operation and shutdown. Central to these was the engine control valve. It accepted electrical signals for engine start and stop, and through the use of solenoids (electrically operated actuators), it orchestrated the hydraulic pressure that would operate the main propellant valves. It included built-in redundancy to ensure engine shutdown should the primary stop solenoid fail.

Another was the ignition monitor valve which used a Mylar diaphragm to sense the pressure in the thrust chamber. Its task was to allow the main fuel valves to open only when the engine had been positively ignited. The rise in chamber pressure upon ignition would push against this diaphragm. This actuated the poppet within the valve, permitting hydraulic fluid to reach the main fuel valves and open them.

Igniters

An F-1 engine had five igniters, each of which was involved in smoothly and safely starting the engine. Four of these were electrically operated pyrotechnic devices; the fifth was a chemical cartridge that provided the impetus for the main combustion to begin.

Two of the pyrotechnic igniters were used to ignite propellants in the gas generator to initiate the process that would spin-up the turbopump. The other two pyrotechnic igniters were installed

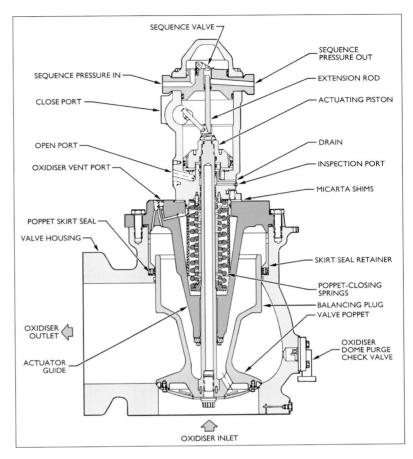

ABOVE Cross-sectional diagram of the F-1 engine's main LOX valve.
(NASA/Woods)

in the nozzle extension so that when fuel-rich turbine gas entered the nozzle, it would immediately begin to burn with the LOX that would already be present at that time.

Each of these igniters contained a small squib that was fired by applying an electrical current at 500V across it. A diode in the circuit ensured that any voltage below 250V didn't set it off. This squib then ignited a larger main charge that would burn for up to 9½ seconds, during which time it would initiate combustion at whichever site it had been installed. Embedded within the main charge was a piece of copper wire which formed an electrical link. This was burned away within the first second of operation and it therefore provided a clear indication to the engine's control system that the igniter was definitely operating.

Main combustion in the thrust chamber was assured by employing a hypergolic chemical reaction. Hypergols are chemicals that spontaneously ignite when mixed. In this case, a fluid comprising 85% triethylborane and 15% triethylaluminium would ignite when mixed with

ABOVE An F-1 engine is tested at the T-Stand at MSFC. *(NASA-MSFC)*

LOX. The fluid was stored in what NASA termed a *hypergol cartridge* that had been installed in a special manifold. The ends of the tubular cartridge were diaphragms designed to burst at a specific pressure.

After fuel from the turbopump had been introduced to the manifold, its rising pressure would cause the two disks to burst. The fuel flushed the hypergolic fluid out of the cartridge and into the thrust chamber via 25 orifices built into the face of the injector. There the mixture of fuel and hypergolic fluid would meet and mix with the LOX already being injected into the chamber. The reaction of the hypergolic fluid with the LOX ensured ignition across the face of the injector. This method of ignition continues to be used for kerosene/LOX engines like the Merlin used on SpaceX's Falcon launch vehicle.

Two valves were closely associated with the hypergol cartridge and therefore had important roles in starting the engine. The ignition fuel valve kept the hypergol manifold closed off until fuel pressure had risen sufficiently. As this valve opened, high pressure fuel would burst the cartridge disks and transport the hypergolic fluid into the chamber to ignite the propellants. When this occurred it also unlocked the ignition monitor valve whose Mylar membrane sensed a rise in pressure in the thrust chamber. This then ensured that hydraulic fluid would open the main fuel valves only when initial combustion from the arrival of the hypergolic fluid had been sensed.

Thermal protection

For most of the time that the F-1 burned, it was operating in a near vacuum and flying faster than the speed of sound. Unlike at sea level, when the exhaust was contained by air pressure into a narrow stream, at altitude the plume expanded markedly as it left the nozzle.

Moreover, turbulence and flow separation around the base of the S-IC generated a backflow that bathed the lower part of the rocket in hot exhaust gas. Images from tracking cameras show flames reaching as far up the S-IC as the intertank section. As a result, the engines required to be protected from the hot gases that surrounded them on ascent, and from the infrared radiation from the exhaust plume.

Two types of thermal protection were installed on the engine prior to launch. Large portions were covered with foil batting which comprised two sheets of Inconel foil ranging from 0.1 to 0.15mm thick that sandwiched heat resistant fibre. Selected areas were protected by layers of asbestos that were reinforced with Inconel wire and faced with aluminium foil. A series of fittings were added around the engine, including studs welded to the reinforcement bands around the thrust chamber. The insulation was attached to these fittings. Preformed brackets allowed the batting to stand off from the complex machinery within.

Starting the F-1

Bringing this enormously powerful behemoth of an engine to life was a complex and carefully orchestrated sequence of events. Like musical notes, they had to occur at the right time and in the correct order. After commanding the engines to start, 8.9 seconds were allotted to bring all five F-1s up to full thrust before allowing

ABOVE An F-1 engine mounted in a test stand showing the thermal protection system; foil batting that would protect it from hot backflow gases during flight. *(NASA/Rocketdyne/Mike Jetzer)*

BELOW Foil batting partly added to a nozzle extension. *(NASA/Mike Jetzer)*

BELOW This engine includes brackets that were fitted to support the thermal protection system batting. *(Mike Jetzer-heroicrelics.org)*

them to lift 3,000 metric tons of Saturn V and its payload off the launch pad and to initiate the next movement in an Apollo symphony.

Upon receiving the command to begin at 8.9 seconds before lift-off, the ignition sequence's opening notes were a set of initial preparations. A valve moved to divert used hydraulic fluid back to the engine rather than to the equipment on the launch pad. Heaters that surrounded the bearings at the cold end of the turbopump were turned off. Then 500V was applied to all four pyrotechnic igniters – two in the nozzle extension and two in the gas generator. These would burn until the engine was at full power.

The next bar in this technical prelude began when copper wire links within the igniters were burned away. This yielded an electrical signal that indicated they had successfully fired and it was a cue for a *Start* command to be sent to the engine control valve. The rising music of the F-1's ignition began to pick up pace.

The engine control valve rerouted the hydraulic circuit to have it become partially self-contained within the engine. It passed hydraulic fluid to the *Open* port of the two main LOX valves. These started to open, allowing huge quantities of the extremely cold oxidiser to pass through the LOX pump into the LOX dome, heading for the thrust chamber. This flow acted on the pump's inducer and impeller as if they were a turbine, causing the shaft to slowly turn and initiate the operation of the turbopump.

As the LOX valves reached 16.4% of their travel, the pins at the end of their actuating rods began to open their associated sequence valves, allowing hydraulic fluid through to open the ball valve of the gas generator. Both propellants sprayed into the gas generator's combustion chamber where, thanks to the flaming igniters still alight, they began to burn and produce copious amounts of exhaust gas.

The fuel-rich exhaust from the gas generator entered the turbine housing and was directed onto the blades around the outside of the two turbine wheels. Now the great shaft that ran along the length of the turbopump began to rotate faster, turning the pumps and raising the pressure of the propellants on their way to the main engine valves. In a harmonious counterpoint to the F-1's opening music, the turbine exhaust began to heat the coils in the heat exchanger, its warmth producing gases that would pressurise the propellant tanks. Then like air moving along the passages of a great tuba, the exhaust entered the wraparound manifold which ushered it into the nozzle extension. Still loaded with unburnt fuel, it met the LOX that was pouring down from the injector, where the flames from the igniters caused it to combust.

The main refrain from the accelerating turbopump continued. As it gathered speed and rose to a scream, pressure in the propellant lines rose dramatically. The rising fuel pressure opened a valve to inject fuel onto the shaft's bearings for lubrication. At the manifold where the hypergol cartridge sat, the fuel pressure opened another valve, the ignition valve, which permitted fuel to pass to the cartridge. In quick succession, the diaphragms at each end of the cartridge burst and a pulse of fuel laced with hypergolic fluid sped to the injector along dedicated lines. When it met the LOX within the chamber, the vigorous reaction ensured combustion across the face of the injector.

Now the ignition monitor valve picked up the tune. The bursting of the hypergol cartridge caused a spring-loaded rod to unlock this valve. Its job was to verify that combustion had successfully begun. The sudden rise in the chamber's pressure acted on the valve's Mylar diaphragm, thereby operating its poppet.

With combustion assured and LOX successfully flowing into the thrust chamber, the final crescendo of the F-1's overture could play out. With only three seconds or less remaining to the scheduled lift-off, the operation of the ignition monitor valve permitted hydraulic fluid to flow to the *Open* ports of both main fuel valves, actuating their poppets. As the fuel valves opened, fuel entered both inlets of the fuel manifold above the injector. Two-thirds was diverted through the pipework that formed the walls of the thrust chamber. These pipes had been preloaded with ethylene glycol and this was flushed out by the fuel. Consequently, the initial dose of fuel to the injector was heavily loaded with glycol that moderated the rise in the engine's thrust and lessened the resulting shock on the rocket's structure.

It took only about one second for the main fuel valves to open and the F-1 to reach its

ABOVE **Five F-1 engines emit their final gasps as S-IC-6 departs from the Apollo 11 vehicle.** *(NASA)*

rated 6.7MN sea-level thrust. A sensor detected when it had reached 90% of this value and this was signalled to the rest of the system. For a second or two (depending on the staggered sequence of engine starts) all five F-1s at the base of the Saturn V belched their full fury into the 12m deep flame trench over which the rocket sat. There, a huge wedge-shaped flame deflector forced the fire to either side. When all five engines were confirmed to be operating satisfactorily, the hold-down arms that clamped the Saturn in position were let go and the rocket flew free at last.

The F-1s on the S-IC burned for about 2½ minutes, limited by the tonnage of propellant that could be held in the enormous 10m tanks above. As the vehicle ascended, the departing exhaust gases encountered a rapidly thinning atmosphere, allowing them to exit the engines at greater speed and improving the efficiency of the engines by up to 20%. An engine whose thrust began at 6.86MN (1,542,000lb-f) was pushing the vehicle with a thrust of 8MN (1,800,000lb-f) by the time it was shut down. As the tanks drained and the stage lightened, the acceleration rapidly increased. The central engine was shut down first to limit the acceleration in the final phase. When sensors detected low propellant levels and shut down the other engines, the vehicle was flying at 8,650km/hr with respect to the ground. It was 66km in altitude and had travelled 93km downrange (Apollo 11 figures).

Shutting down the F-1

Shutdown of an F-1 was not left to chance. During the engine's test phase, it was important to terminate combustion in a fashion that minimised damage to its systems. In flight, an orderly and controlled shutdown was necessary to make the separation of the Saturn's first and second stages as predictable as possible. Had the tanks been allowed to run dry, it was possible that in their dying moments, the engines could burn very fuel-rich or LOX-rich in ways that were unpredictable and damaging. More importantly, if they were not shut down together they might impart an unwanted rotation on the stage.

Shutdown was achieved by applying an electrical *Stop* signal to another solenoid on the engine control valve. This sent hydraulic pressure to the *Close* port of the propellant valves, including the gas generator's ball valve. The pressure was applied in such a manner as to cause these valves to close in sequence: gas generator ball valve first, then the LOX valves and finally the fuel valves. With the gas generator starved of propellants, the supply of turbine gas faded and the turbopump began to spin down. This caused the fuel and LOX pressures to plummet and thereby close those valves that were held open by that pressure, particularly the igniter fuel valve. At the same time, heaters at the cold end of the turbopump shaft were turned on to avoid condensation.

At shutdown, the engine's thrust didn't reduce instantly but rather exhibited a characteristic tail-off that took 0.7 seconds to reach negligible values. During that time, it managed to add an extra 10m/sec to the velocity. This tail-off thrust had to be accounted for when timing the separation of the stages and deciding how much additional rocketry would have to be carried aboard the various stages to pull them apart.

The F-1 in context

Today, the F-1 engine remains the most powerful single-chamber liquid-fuelled rocket engine ever developed. Despite the difficulties that had to be overcome during its tortuous development, it gained an enviable service record and flew on thirteen successful Saturn V launches. Using a different basic design, the Soviet Union did develop a more powerful and more efficient engine, the RD-170 and its RD-171 variant. It was used on the Energia vehicle that flew twice, one of which lofted the Buran space shuttle on its only flight.

On the RD-170/171, the problem of combustion instability was overcome by having four smaller combustion chambers as opposed to the F-1's single large chamber. It could be argued that having thirteen compartments

BELOW A complete F-1 engine without its nozzle extension. *(NASA-MSFC)*

between the baffles of an F-1 injector went a little way towards giving the engine separate discrete chambers.

The increased efficiency of the RD-170/171 was achieved by using staged combustion whereby, instead of dumping inefficiently burned turbine gases near the end of the nozzle's expansion, they were injected into the chamber to fully complete their combustion. At 7.9MN thrust, the pressure within its chamber was also more than three times higher than that of the F-1.

An uprated version of the F-1 was developed by Rocketdyne in the mid-1960s. Known as the F-1A, it was very similar to the F-1 but benefitted from detailed revision and strengthening of the engine's components. In particular, the turbopump used a smaller but faster-spinning redesigned turbine that ran at 75,700bhp to generate higher pressures in the propellant lines. This raised the engine's sea-level thrust from 6.77MN of the later F-1s to 8MN.

Two F-1 engines were modified to test aspects of the F-1A design; for example, an altered start sequence. However, by the time an engine was being built to full F-1A specification, NASA's budget was in decline and the programme was cut. It only ever reached 40% completion and would have required another ten months to prepare it for testing. The first generation F-1s were sufficient for all upcoming missions.

A generation later, thought was given to reviving and redesigning the F-1 for the space initiatives that were being considered by successive US administrations. Such engines would have benefitted from advances in metallurgy and computer-aided techniques to redesign and refine their systems, but scant funding prevented working units from being developed.

Fishing for the F-1

All thirteen S-IC stages that began the flights of the Saturn Vs were allowed to crash into the Atlantic Ocean 650km from the Kennedy Space Center, where they sank in over 4,000m of water.

In 2011, wealthy entrepreneur Jeff Bezos initiated a project to search the floor of the ocean for F-1 engines and bring examples to the surface for conservation and display. Sufficient components had been tracked down by mid-2013 for two complete displays. Using remotely operated submersibles, his team lifted these parts to the surface. On close examination, the serial number 2044 was found on one of the thrust chambers. It confirmed that they had recovered the centre engine from S-IC-6, the first stage that lifted Apollo 11 on its historic mission, a truly lucky find.

BELOW The twisted remains of an F-1's upper thrust chamber found resting on the sea floor more than 4km deep. *(Courtesy of Bezos Expeditions)*

BOTTOM With its copper face green with corrosion, this F-1 injector plate is given a rinse with fresh water to remove salt after being recovered from more than 4km deep in 2013. *(Courtesy of Bezos Expeditions)*

Chapter Three

S-IC: The monster stage

The power to lift 3,000 tons of space vehicle off the ground and send it well on its way to orbit came from the mighty S-IC stage, a machine that was little more than two giant tanks feeding five huge F-1 engines. Yet, for about 160 seconds, this stage could develop power levels approaching 60 gigawatts, roughly as much power as the United Kingdom's entire electrical generating system produces at peak time.

OPPOSITE **S-IC-11 lifts Apollo 16 off the launch pad on 16 April 1972.** *(NASA)*

RIGHT The interior of the Michoud Assembly Facility, a cavernous space under cover where Boeing built the S-IC stage. Four partially completed stages are lined up at the far side while the foreground is populated with conical fairings. *(NASA-MSFC)*

The contract to build the huge S-IC stage was awarded to Boeing on 15 December 1961. Boeing staff quickly moved into MSFC to work with NASA's engineers on the design of the stage and to plan for its manufacture. The company already had much experience building large aircraft for the military and civilian markets, but the sheer scale of the stage taxed the skills of its people.

Starting in early 1964, two test stages and the first two flight examples were built at MSFC. Then once NASA was satisfied that major bugs in the S-IC's fabrication had been ironed out, a further fourteen stages, including a dynamic test article, were produced at the Michoud Assembly Facility near New Orleans. Though some S-IC testing occurred at MSFC, most stages were shipped 50km from Michoud to the Mississippi Test Facility to be tested as a unit in one of the massive test stands installed there in support of the Apollo programme. They were eventually barged to the Kennedy Space Center to be integrated into a Saturn V vehicle within NASA's iconic Vehicle Assembly Building.

In most respects, the S-IC was a conservative design. Where it did push the boundaries was by its sheer size, demanding innovations in welding techniques to construct its two gargantuan tanks. Its propellants,

LEFT S-IC-D, a test version of the S-IC stage specifically built for dynamic testing, is lifted into the dynamic test stand at the Marshall Space Flight Center. The stage is outfitted with four weight simulators and one dummy engine instead of working versions of the F-1. *(NASA-MSFC)*

ABOVE An S-IC is carried aboard the barge *Pearl River* from the Mississippi Test Facility to the Michoud Assembly Facility. *(NASA-MSFC)*

RIGHT Diagram of the S-IC stage indicating its major components. *(NASA/Mike Jetzer/Woods)*

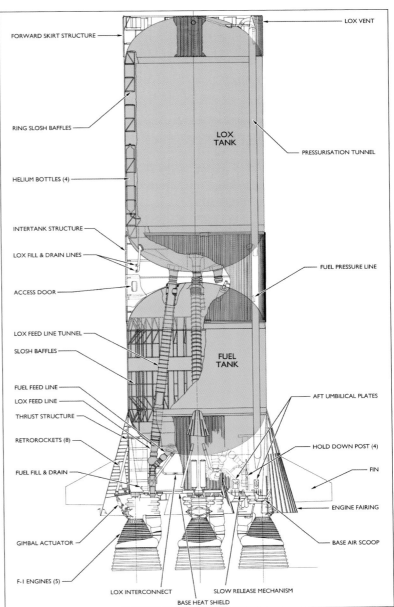

kerosene (RP-1) and liquid oxygen (LOX) were well understood, having been used for the Jupiter, Thor and Atlas vehicles and for the Saturn S-I stage. However, the thirsty F-1 engines took rocket engineering to a far higher power level; in this case a total stage output of 33.85MN (7,610,000lb-f).

Thrust structure

The S-IC consisted of five major components. At the bottom was the thrust structure comprised of two strong rings, the lower of which supported the four outer engines. A fifth engine was mounted in the middle of the ring at the centre of a cruciform of I-beams.

Eight longitudinal structural members connected the rings. Four were thrust columns directly above the engine attachment points. These spread the force of the four outer engines across the structure. Four hold-down posts between the thrust columns led from points where the weight of the whole vehicle was supported on the launch pad. Thus the thrust structure acted as a structural interface between the eight places on the lower ring where forces were concentrated in small points, and the cylindrical skin of the fuel tank where the same forces had to be evenly distributed.

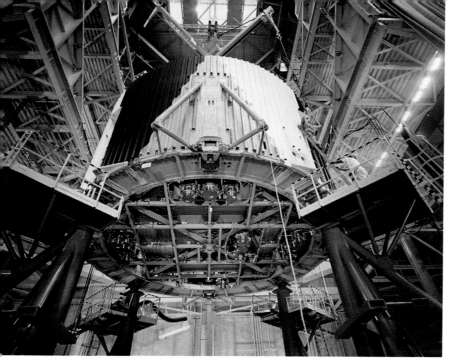

ABOVE An S-IC thrust structure is mounted on four support columns at the Michoud Assembly Facility ready for the rest of the stage to be assembled above. This view from below shows the underside of the structure and includes the engine mounts on the lower ring and the ends of the propellant ducts. *(NASA-MSFC)*

BELOW Drawing of the thrust structure including engine fairing and fin. *(NASA)*

BELOW RIGHT Drawing of a tail service mast. This was connected to the thrust structure until lift-off. As the Saturn V began to ascend, the mast disconnected and rotated into its housing to be protected by a hood that came down automatically. *(NASA)*

As the rocket moved through the atmosphere, a turbulent backflow of exhaust bathed the rear of the vehicle in hot gases. A large heatshield was built across the base to protect the propellant ducts and other internal elements. The skin panels that clad the structure were made from 7075-T6 aluminium, an alloy which includes about 5% zinc as well as magnesium and copper. These panels were given reinforcement with longitudinal stringers of a rectangular cross section added to their outside face.

Four conical fairings were attached around the outside of the structure to shield the nozzles of the four outboard engines from the airflow. Without this protection, aerodynamic forces would interfere with the gimbal actuators that swivelled the engines to steer the vehicle.

Each fairing also sported a large fin, four in all, each faced with titanium. These fulfilled the same role on the Saturn V as does the empennage of an aircraft (its tail plane and vertical stabiliser); that of providing aerodynamic stability, much like the feathers of an arrow or dart. Being fixed, the fins played no role in steering the rocket – this was achieved solely by gimballing the four outer engines.

The thrust structure included fittings that interfaced the stage to its launch platform. Umbilical connections were made via tail service masts mounted on the launch platform. These were to fill and drain the fuel tank and to service

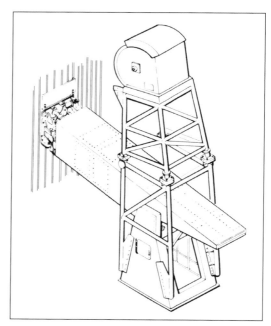

RIGHT Cutaway drawing of a hold-down arm. The pincer mechanism is shown in its hold-down state in green. The dashed outline shows it in its released state. *(NASA/Woods)*

CENTRE Detail drawing of the controlled release mechanism. *(NASA/Woods)*

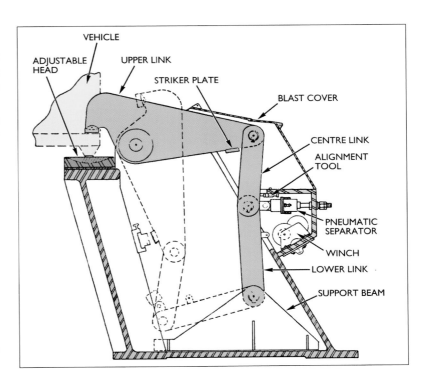

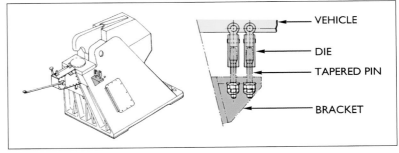

the engines prior to launch with power and ramjet fuel. The fuel was supplied to lubricate the engines in lieu of the RP-1 that would be used during flight.

The launch platform included a large square hole, 13.7m on each side, which accommodated the engine bells and their exhaust. Mounted around the edges of this hole were four strong supports called hold-down arms upon which the entire weight of the space vehicle sat. The lower ring of the thrust structure had four hold-down points to accommodate the supports. Clamps on the hold-down arms held the hold-down points like pincers to hold the vehicle firmly in place throughout the weeks of preparation in the VAB, during transportation to the pad, and on the pad right up to the moment of launch. When lift-off time arrived and the thrust was verified on all five engines, the mechanism that held each arm to the vehicle was pneumatically collapsed, thereby releasing the Saturn V.

A series of fittings near the hold-down points allowed a number of dies to be attached to the thrust structure for what were called *controlled release mechanisms*. When the hold-down arms released, the rocket was not allowed to just leap off the pad. To lessen the shock of release, the initial 15cm of the rocket's ascent was impeded by a series of tapered rods that had to be pulled through these dies by the rising stage. The rods were attached to fittings on the hold-down arms.

Fuel tank

Above the thrust structure was the fuel tank which held about 820,000 litres of RP-1, an amount that increased slightly through the programme. Although this fuel is less efficient

RIGHT An S-IC fuel tank at MSFC for pressurisation tests. *(NASA-MSFC)*

RIGHT Interior of an S-IC fuel tank. One of the ducts that carried LOX through the fuel tank is on the left. A series of anti-slosh baffles were mounted around the inside of the tank's walls. *(Courtesy of US Space & Rocket Center)*

than hydrogen it has a number of advantages. It is a liquid at normal temperatures, it stores a lot of energy in a given volume, hence smaller and lighter tankage, and it can be easily handled in large quantities. And of course it could provide the sheer power that was needed in the first stage to get the whole vehicle moving.

The fuel tank was built in three major sections using a type of aluminium alloy known as 2219-T87 that contains about 6% copper. It is resistant to stress corrosion cracking (where the damage caused by corrosion combines with mechanical stress to promote the growth of cracks) and it is relatively easy to weld. A cylindrical wall, 4.9mm thick at the bottom and thinning to 4.3mm at the top, was capped by ellipsoidal bulkheads to form the 13.13 × 10m component.

This wall was made from a series of plates, each of which was milled from a flat plate to include vertical stringers for stiffness. Once milled, the plates were given the required curve before being heat treated and welded into a cylinder. A series of nine rings, or baffles, like giant washers ran around the internal circumference of the tank to reduce any tendency of the propellant to slosh. These were attached to the internal edges of the stringers.

The bulkheads were constructed from eight large gores like slices of a pie, each made from two sheets; a base segment and an apex segment. These were milled to be thinner towards the apex while keeping a sufficient thickness at the edges for welding. The milled flat sheets were then bulge formed into the required shape, a process that used hydraulic pressure to force the sheet into a die. Heat treatment and welding then yielded a full-size bulkhead. Rings with a 'Y' cross section were welded between the cylindrical wall and the

CENTRE An ellipsoid bulkhead for an S-IC fuel tank mounted on a jig for making the radial welds that join the individual gore segments. *(NASA-MSFC)*

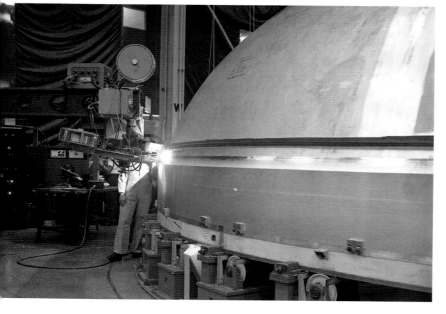

LEFT A fuel tank bulkhead is welded to a Y-ring to permit its subsequent attachment to the wall cylinder below, and to allow it to be affixed to the structural piece above. *(NASA-MSFC)*

bulkhead. These also provided a means to attach the tank to the adjoining sections.

Since the LOX tank was at the top of the stage, the ducts that led from the tank to the engines had to somehow get past the fuel tank. On the S-II and S-IVB stages the ducts from the upper tank were routed around the outside of the lower tank. However, the F-1 engines consumed propellant at such a prodigious rate that designers wished to avoid introducing bends in these ducts because a kink in the line would reduce the feed pressure and risk cavitation at the inlet to the engine's turbopump.

Designers chose instead to create five large straight tunnels, 63.5cm in diameter, which ran right through the fuel tank, piercing both its top and bottom bulkheads. Each tunnel contained a 50cm duct through which the extremely cold oxidiser passed directly from the LOX tank down to each engine. The gap between the tunnel and the duct provided a degree of insulation between the RP-1 outside the duct and the extremely cold LOX within; a temperature difference of about 200°C.

RP-1 from the fuel tank reached the engines via ten short 30cm diameter ducts, two per engine, that led from the tank's aft bulkhead. As the outlets to these ducts were arranged in a ring, a portion of the tank's base was lower than the outlets. Because any fuel in this region would be unable to reach the engines this volume was occupied by a foam insert.

All 15 ducts (10 for fuel and 5 for LOX) included prevalves that prevented the propellant from entering the engines until required. They also provided a backup method of halting combustion in case of failure of the engines' main valves.

Intertank structure

Both the fuel and LOX tanks had outward-facing domed bulkheads. To keep them apart, a spacer known as the intertank was required. This cylindrical component, 10m in diameter and 6.66m tall, was constructed using a series of rings as supports for attached skin panels, all made from 7075-T6 aluminium.

The skin panels were given their distinctive appearance by being corrugated. Panels were included that allowed connection between the stage and the ground equipment to fill and drain the LOX tank and to vent the fuel tank. There was also an access door.

LOX tank

The largest single component of the S-IC was the LOX tank. Measuring 19.53 × 10m, this vessel held 1,340,000 litres and was essentially a stretched version of the fuel tank. The mass of liquid oxygen in this tank, nearly 1,500 metric tons, represented just about half of the mass of the entire vehicle.

Two bulkheads, domed in an ellipsoidal form, capped either end of a large cylindrical section, all of which was built from 2219-T87 aluminium alloy. The walls thinned from 6.5mm at the bottom to 4.8mm at the top and there were

ABOVE LOX tunnels within an S-IC fuel tank. Ducts within these tunnels brought LOX directly through the tank. The outlets for fuel to pass to the engines, two per engine, formed a ring near the centre of the bulkhead. *(NASA/Mike Jetzer)*

LEFT The LOX tank from S-IC-S, the structural test version of the stage, seen here at MSFC. The forward skirt has been attached at the top. *(NASA-MSFC)*

stiffening stringers machined as part of the skin. Fifteen anti-slosh rings were fitted around the internal circumference.

Four 60cm diameter aluminium bottles, each 5.5m long, were installed on one side of the tank and supported by the anti-slosh baffles. They held helium gas that would be used to pressurise the fuel tank. The extremely cold conditions within the LOX tank suited the storage of helium because it allowed more of the gas to be held within the design pressure of the bottles than was possible at ambient temperatures.

Although the LOX tank contained a huge quantity of liquid oxygen at -183°C, it was not insulated. Instead, the loading of the tank before a mission caused ice to form on the exterior surface by the condensation of water from the humid Florida air. This ice provided a degree of insulation. Yet as heat leaked in through the skin, it caused the LOX to boil and vaporise, and the resulting gas was vented. This loss was compensated by constantly pumping in more LOX until the point at which venting was terminated and the tank was pressurised. As soon as the engines ignited and the stack began to vibrate, the accumulated ice came away in an avalanche of white crystals and chunks.

Five outlets at the base of the tank led to five 50cm ducts that passed through the fuel tank to each engine. Each outlet included a device to inhibit vortices. A large cruciform baffle at the base of the tank helped to minimise slosh and further reduce vortexing as the tank neared depletion.

LEFT Interior of LOX tank looking towards the bottom. Circumferential anti-slosh baffles ring the tank. The technicians are standing near four large 'bottles' used to store helium at cold temperatures. During flight, the helium stored in these bottles was warmed in the engines and used to pressurise the fuel tank. *(Boeing/NASA)*

Forward skirt

Another large ring topped off the stage. It provided clearance for the domed upper bulkhead of the LOX tank and supported the S-II, S-IVB and payloads above. This so-called 'skirt' was of similar construction to the intertank structure, but at 3.07m tall it was less than half the height because it had to accommodate only a single dome.

The forward skirt housed some of the stage's electrical and electronic hardware, including range safety and telemetry equipment, along with a system that controlled the temperatures of these units. Prior to propellant loading, cooled air was supplied. Then as LOX began to be loaded into the giant tank below, warmed nitrogen gas was fed into the equipment's housing. Later, as the five cryogenic J-2 engines on the S-II stage above were chilled down, the supplied gas was made even warmer in an effort to maintain the equipment's internal temperature.

Pogo suppression

When the unmanned Apollo 4 flew in November 1967, objectionable pogo (see box overleaf) didn't make an appearance on what was the first flight of the S-IC. In this case, the vehicle was lightly loaded. But pogo did surface on the second unmanned flight, Apollo 6.

This mission had included a dummy load to represent the lunar module, an addition that changed the Saturn V's inherent resonant frequency. For a 30-second period towards the end of the S-IC's flight, pogo vibrations of ±0.65g at a frequency of about 5.3Hz were measured at the spacecraft. Had there been a human crew, such violent shaking would have been hugely uncomfortable if not physically

ABOVE A forward skirt is being lowered at MSFC, to be fastened to the Y-ring at the top of an S-IC LOX tank. *(NASA-MSFC)*

LEFT This set of five images was taken across three months as an S-IC stage was assembled from its five component parts; thrust structure, fuel tank, intertank, LOX tank and forward skirt. *(NASA/Mike Jetzer)*

> ### SHAKING ALL OVER: THE POGO PHENOMENON
>
> One of the biggest headaches that faced designers of the early liquid-fuelled rockets was a tendency for their devices to exhibit lengthways vibrations. The phenomenon came to be known as 'pogo' after a children's toy which consisted of a telescoping stick with a spring that enabled its user to bounce along. The child would use their legs and arms to put energy into the stick's spring at the right time to keep the bounce going. Pogo in rocketry treats the entire vehicle as one long spring with its own mechanical resonant frequency. If the vibrations that result from the pogo effect are allowed to become excessive, they can be damaging to the rocket or its payload and can easily shake a human crew senseless.
>
> Energy for pogo vibrations comes from an unfortunate interplay between an engine's thrust and the movement of large volumes of liquid down long ducts. Both of these systems resonate and it works something like this. All engines exhibit minor variations in their thrust across short timescales. A small peak in the thrust accelerates the vehicle slightly and also causes it to compress a little. Two things then occur. There will be a resultant short-term rise in the pressure of the propellants at the inlets to the engine's pumps which will feed through to cause another positive bump in the thrust, thereby giving the engine an inherent resonant frequency. Also, the vehicle will rebound from its compression, causing a negative bump in the propellant pressure and likewise on the thrust. If the vehicle includes long feedlines, these too have resonant frequencies.
>
> If these resonant frequencies happen to coincide, a coupling effect can occur where the vibrations occur in sympathy and build in such a way that they reinforce each other. This can result in severe pogo. An additional problem with a liquid-fuelled rocket is that as the tanks empty, the resonant frequency of the vehicle changes, sweeping through a wide range over the course of a stage's burn. Also, the pattern of resonance will change depending upon variables like the mass of the payload. This makes it difficult to predict this effect and therefore counter it structurally.

Gases can be compressed whereas liquids are virtually incompressible. As a result the gas-filled voids acted like springs. The springiness of the gas altered the resonant frequencies of the LOX ducts, lowering them and making it far less likely that they would match the inherent resonant frequencies of the engines. This hugely reduced the tendency for the two systems to vibrate in sympathy and thereby limited the size of any pogo oscillations.

Prelaunch fuelling

Loading fuel into the S-IC was relatively straightforward because RP-1 is a stable liquid at room temperature. This task occurred about three weeks prior to launch and took about two hours.

Fuel entered the tank via a 15cm duct near the bottom at a rate of 6 metric tons per minute until it was 98% full, at which point the flow was reduced to 0.6 metric tons per minute until it was 2% above the expected amount. With one hour remaining to launch, a small amount was drained from the tank until the required fuel load had been achieved.

The quantity of fuel in the tank was measured using a series of probes that reported the level in the tank using electrical capacitance. Additionally, the temperature of the fuel was measured so that the density of the liquid could be determined. In order to maintain an even density throughout the tank, nitrogen gas was injected into the ducts leading to the engines and allowed to bubble up through the fuel, gently stirring it in the hours leading up to launch.

harmful. This is especially so because it would have occurred at the same time that the vehicle's acceleration was heading past 3*g*.

To cure the S-IC pogo, engineers altered the one thing they could; the resonant frequency of the LOX ducts, and they managed to do so without adding much mass to the vehicle. At the bottom of each duct was a prevalve that kept the oxidiser from entering the engine until required. By virtue of their design, these included cavities. Engineers modified the prevalves for the four outer engines to allow these cavities to be filled with helium just prior to flight.

LOX loading

Whereas fuel loading didn't present major problems, LOX loading required a lot more care. For a start, many substances are rendered extremely flammable, almost explosive, when they come into contact with LOX. Therefore, it was important that the interior of the massive LOX tank be made spotlessly clean prior to loading. Indeed, the aluminium alloy plates from which it was made were carefully degreased and washed prior to fabrication. This care over its cleanliness was maintained throughout the manufacture and test processes up to the time when the tank was filled.

One technique that was used to keep the tank clean was to have it slightly pressurised with nitrogen during storage and transportation. Maintaining a pressure about 0.3bar above atmospheric not only kept contamination out, it also helped the tank to retain its shape.

Nine hours prior to launch, the LOX tank was given a final purge with nitrogen gas to ensure that contaminants like water vapour were expelled. Otherwise water vapour could condense and form ice particles as hard as rock which could be ingested by the engines.

The loading process began by piping in LOX at 95 litres per second, a relatively slow pace given the enormous size of the tank. As soon as this -183°C liquid met the tank walls, it boiled furiously as it vaporised. It was like adding cold water to a dry saucepan that had been on the stove too long. Nevertheless, cold liquid continued to be added and gradually the temperature of the structure fell until more and more of the LOX remained as a liquid at the bottom.

When the level in the tank reached 6.5%, LOX began to be pumped in at over 600 litres per second. For 45 minutes or more, the level rose and the warm walls continued to cause boil-off. By the time the level had reached 98%, over a quarter of the LOX that had entered the tank had boiled off and been vented as vapour.

The slow fill rate was then re-established to bring the tank to 100%. As heat entered the tank from outside during the five hours remaining to launch, the level was topped up to compensate for boil-off.

Bubbling

A loaded LOX tank meant that the five ducts that led through the fuel tank to the engines were also filled with the supercold fluid. Although an insulating gap had been created between the ducts and the tunnels that housed them, sufficient heat passed from the warm kerosene that, in the absence of intervention, the ducts could begin to geyser as great bubbles of oxygen gas rose up. In addition, it was essential that the LOX near the engines be cold enough not to vaporise too readily as it entered the engine during start-up.

During the period leading up to launch, engineers instigated a convective flow to keep the LOX moving within the ducts. This inhibited excessive vaporisation and maintained the presence of cold LOX near the engines. It was achieved by injecting a supply of helium gas into the bottom of ducts 1 and 3 and opening valves within a set of interconnects that linked the bottom of the ducts.

As the helium bubbled up ducts 1 and 3 it created an upward flow of LOX there which induced a downward flow within ducts 2, 4 and 5 via the interconnecting pipes. Once this convection was established, the flow would sustain itself until the interconnect valves were closed with only a minute to go to launch.

Tank pressurisation

The fuel tank was pressurised with helium which, prior to lift-off, was supplied from the ground. For flight, the helium supply was switched to the four large storage bottles installed in the LOX tank. As the flight progressed, the pressure in the helium bottles would decrease.

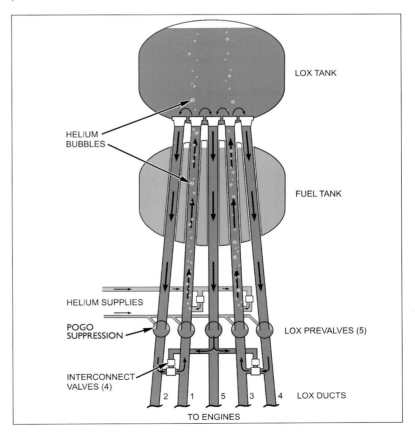

BELOW Prior to launch, cavities in the LOX prevalves were filled with helium for pogo suppression. Helium bubbles set up convection currents within the LOX ducts to inhibit vaporisation and geysering. The bubbles were introduced into ducts 1 and 3, creating an upward flow. Cold LOX then descended ducts 2, 4 and 5. *(Woods)*

A series of five valves arranged in parallel would then be opened consecutively in order to allow more of the gas through to the fuel tank.

The cold helium gas was routed to all five F-1 engines, where it passed through heat exchangers to be warmed by their turbine exhaust. Hot and expanded helium gas was then routed back to the top of the fuel tank to enter via a distributor.

The LOX tank was also pressurised with ground-sourced helium until lift-off. For flight, it used the LOX itself as a source of pressurising gas. In the engines, a little high-pressure LOX was tapped off and fed to the heat exchangers where the hot turbine exhaust turned it into gaseous oxygen (GOX). This was fed to the top of the LOX tank via a venturi (a deliberate narrowing of the feedline) which controlled its flowrate and thereby the pressure in the tank. Calibrated orifices at the inlets to the heat exchangers of the F-1s helped to limit the incoming pressures to the coils within.

Level sensing

In common with the other stages of the Saturn V, a combination of point level sensors and full length sensors based on the capacitance of a tube-within-a-tube (see page 95) were used to determine the amount of propellant in the tanks. The main task for the full length sensors was to help control the propellant loading process and achieve the desired propellant quantity at launch.

Shutting down the S-IC stage was effected by a set of point level sensors mounted near the top of each of the five long LOX ducts. These sensors consisted of glass prisms that reflected a beam of light internally if they weren't surrounded by liquid. The reflected beam could be detected and used to derive a signal indicating that the sensor was no longer submerged. Immersing the prism in liquid removed the total internal reflection effect because the refractive index of the prism was similar to that of the liquid.

With the centre engine having been shut down early to limit the rocket's acceleration, the task of the level sensors was to ensure that the remaining four engines were shut down cleanly before the LOX was fully depleted.

Range safety ordnance

As was required by the US Air Force, which was responsible for public safety for all the launches from the Cape Canaveral area and the range of ocean to the east, the S-IC had provision to allow it to be destroyed in flight if it were to stray excessively from its planned trajectory. The goal was to disperse the propellants into the atmosphere while the vehicle was still in flight in order to avoid a heavy stage with essentially full tanks impacting populated areas.

This would be achieved by nine 2.2m strips of explosive known as flexible linear shaped charges. Each strip was a length of lead with a V-shaped groove filled with explosive. The groove directed the shock wave from detonation at the skin, vaporising it to cut longitudinal openings in the tank walls upon command from the Range Safety Officer via redundant radio command systems. Three strips were mounted on the fuel tank on one side of the stage and six strips were mounted on the LOX tank on the opposite side. This was to minimise the mixing of the propellants after the tanks had been opened.

Retrorockets

Another example of ordnance on the S-IC were the stage's retrorockets. Within each fairing at the bottom of the stage were fittings to attach a pair of upward-facing solid-fuel rockets, eight in total. Their function was to retard the stage after it had been jettisoned or 'staged' from the S-II, and so increase the separation before any unintended rotation could cause recontact.

When the acceleration on the vehicle was sensed to have dropped below 0.5g upon stage shutdown, a command was sent by the instrument unit to fire detonators. These set off *confined detonating fuses* (CDF) that led to each retrorocket where a Pyrogen unit was ignited. This was a pyrotechnic device that shot out a short-duration flame across the inner surface of the rocket's solid propellant.

Each retrorocket was a steel cylinder, 2.2m long × 39cm. It weighed nearly a quarter of a metric ton and provided 337kN (75,800lb-f) thrust for just over half a second. The fairing didn't provide an exit port for the retro nozzles; when they fired, their exhaust simply blew through the aluminium skin at the top of the fairing.

Nearly all of the S-ICs sported eight retrorockets arranged in four pairs. For Apollo 15, engineers reduced the number of units to two pairs, located in opposite fairings at fins A and C, leaving two fairings empty. This modification was one of a number of changes to improve the overall payload capability of the Saturn V.

In the event, as soon as Apollo 15's first stage had been cut away, the reduced retro thrust and a greater than expected tail-off thrust from the F-1s combined to make the separation of the stages less than desirable. Further analysis revealed that if any one of the remaining four rockets failed, there was a small chance that the stages might have recontacted. It was decided that future S-ICs would return to having a full complement of retrorockets.

In both execution and operation, the S-IC was a beast of a machine and a triumph of engineering excellence in the creation and control of enormous forces. To see its huge size or contemplate the bells of the five engines at its aft end was impressive enough. To witness its fury, even more so. Nevertheless, it was only the start of the show that was a Saturn V launch.

The upper stages needed an energy source that was far more efficient, and they would get it in hydrogen.

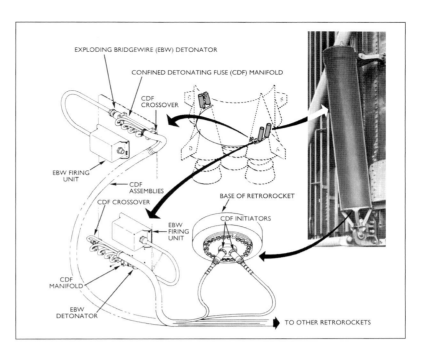

ABOVE Apollo 15 had only four retrorockets whereas other S-IC stages had eight. This drawing shows the arrangement of redundant detonators and fuses that ignited the solid-fuelled rockets. *(NASA/Woods)*

BELOW LEFT An early example of an S-IC (as indicated by the black stripe around its girth) being lifted into the B-2 test stand at the Mississippi Test Facility to be fired as a complete stage. *(NASA-MSFC)*

BELOW S-IC being tested at full power on the B-2 test stand at the Mississippi Test Facility. *(NASA-MSFC)*

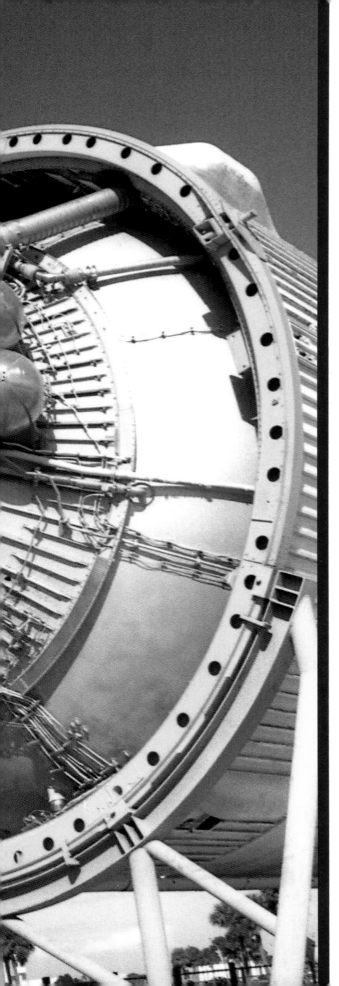

Chapter Four

J-2: Efficiency in an engine

NASA's hydrogen-burning J-2 engine did not possess the brute-force power of the F-1 engine but it was a major contributor to the success of the Saturn V and, like the F-1, it required its designers to up their game by an order of magnitude – and then some. Ultimately, J-2 engines were what really made the Saturn go because they were responsible for most of the velocity given to the spacecraft payload. It was a J-2 that took Apollo out of Earth orbit and towards the desolate landscape of the Moon.

OPPOSITE A J-2 engine at the aft of an S-IVB stage, as photographed by the author in 1994 outside the VAB at the Kennedy Space Center in Florida. This stage is now indoors in the Apollo Saturn V Center nearby. *(Woods)*

Even in the earliest days of theoretical rocketry at the turn of the twentieth century, Russian pioneer Konstantin Tsiolkovsky realised that burning hydrogen in a rocket engine would confer important benefits over most other propellants. However, it would be half a century before hydrogen began to be manufactured in quantities large enough to be used in practical experiments with rocket engines.

The reaction of hydrogen with oxygen is particularly energetic and the exhaust product, water vapour, is environmentally benign. It exits a rocket at extremely high velocities and makes for a particularly efficient engine with a high specific impulse (see page 66) that yields maximum effect for a given mass of propellant.

Immediately after the Second World War, while US Army engineers were learning all they could from the relocated German engineers and captured A-4 rockets, the US Navy had initiated its own studies into rocketry. In association with the Jet Propulsion Laboratory, which was then still an institution that studied 'jet' (i.e. rocket) propulsion, it investigated the possibility of using hydrogen and oxygen to power a vehicle capable of probing high altitudes. In 1947, having brought in the help of a specialist high technology company, Aerojet, these studies led to an engine that produced 13.3kN (3,000lb-f).

Parallel studies were being made by NASA's predecessor, the National Advisory Committee for Aeronautics (NACA). As the world of aviation began to move away from piston engines to jet propulsion in all its flavours, NACA explored many of the technological possibilities, including rocketry. One group, led by a future NASA boss, Abe Silverstein, looked at exotic high-energy combinations like liquid fluorine and hydrogen as well as brainstorming many issues relating to engine design, including the use of showerhead-like injectors.

As Silverstein's group worked towards a 90kN (20,000lb-f) engine in the 1950s, a propellant combination of liquid oxygen and liquid hydrogen (LOX/LH$_2$) was settled upon.

LEFT The RL-10 engine, a hydrogen/oxygen unit developed by Pratt & Whitney. It was the precursor to the J-2 and is still in use today. *(Mike Jetzer-heroicrelics.org)*

To support this research, production of LH_2 was ramped up. With guidance from a military agency established in 1958 named ARPA (Advanced Research Projects Agency), and in association with engine maker Pratt & Whitney, NACA developed an experimental engine, the XLR-115, which would eventually become the workhorse of LH_2 stages, the RL-10.

Centaur

As was often the case during the superheated technical environment that surrounded US rocketry research in the twentieth century, many threads, partnerships and alliances enhanced the continuing development of LH_2 rocketry. As NACA continued their work, the builders of the Atlas booster, Convair, part of General Dynamics, wished to create a powerful upper stage for their rocket and chose LOX/LH_2 as its propellant combination. This would be called Centaur.

Aware of Pratt & Whitney's work with NACA on the XLR-115, Convair chose the RL-10 engine. ARPA ordered six Centaurs, and in doing so, established an evolving line of hydrogen-powered upper stages that continues to this day. Over the decades, these stages have accelerated into deep space such notable craft as the Vikings to Mars, the Voyagers to the outer planets and Cassini to Saturn.

Lessons learned in the development of Centaur and the RL-10 would have a profound influence on the Saturn V upper stages and the J-2 engines that powered them. For example, the Centaur tanks were separated by a common bulkhead that was insulated to prevent the two propellants thermally affecting each other. Although both LOX and LH_2 were extremely cold, the temperature difference across the bulkhead was still nearly 70°C.

The RL-10 itself introduced important innovations to LH_2 rocketry. Its injector face, rather than being made of solid metal, was porous. Layers of steel mesh were sintered (fused together at high temperature and pressure) to form a stiff, yet porous metal block. In use, some hydrogen gas was fed through the pores in the injector to keep it cool while the bulk of the propellants were passed through holes in the injector face.

ABOVE RL-10 engines being worked on at MSFC in 1963. *(NASA-MSFC)*

Another idea was to use the LH_2 fuel as a lubricant for the engine's moving parts. In order to be liquid, hydrogen fuel had to be cooled to only 20K (20°C above absolute zero). It then had to be pumped to the walls of the thrust chamber in a liquid form. Any part of the engine exposed to LH_2 was similarly chilled, including moving machinery that required lubrication. Almost any other substance conventionally used as a lubricant would freeze solid in this environment, but using LH_2 in this role avoided the problem.

The RL-10 drove its turbopumps using a technique known as *closed expander cycle*. Instead of burning some of the propellant in a gas generator, as in the F-1, the RL-10 had its liquid fuel turn to gas within the cooling pipework of the thrust chamber, and this gas then powered the turbopumps on its way to the chamber.

Centaur to Saturn

When NACA was transformed into NASA towards the end of 1958, Abe Silverstein moved into the new agency's upper echelons where he chaired a committee to study the future development of the Saturn vehicle and its

HOW EFFICIENT IS THAT ENGINE?

The most common and easily understood performance number associated with a rocket engine is its *thrust*. Quite simply, this is the force with which it pushes. However, this says nothing about how efficiently the engine can achieve that thrust. An efficient engine will require less propellant to be carried as it works to accelerate the payload.

Another performance figure, and one that addresses efficiency, is *specific impulse* (I_{sp}). But there is scope for confusion there, depending upon whether engineers think about rocket propellant in terms of its weight or its mass.

Browse through the dense technical minutiae of a Saturn V Flight Evaluation Report and you find tables that list many of the performance characteristics of each of the vehicle's eleven main engines. Among them are figures for specific impulse which have been quoted in two ways by the engineers at the Marshall Space Flight Center (MSFC). This is not just a metric versus imperial question. The conceptual approach that leads to each is different.

One of these two quoted figures is a measure of how much momentum is added to a vehicle for each unit mass of propellant consumed. The other figure is a ratio of the engine's thrust to the weight of propellant consumed in a second. The dual approach seen in these documents reflects the influence of the German immigrants at the core of MSFC. In addition to being steeped in the metric system, they conceived of specific impulse in a different way from their American colleagues who preferred imperial units.

For generations, weight and mass were treated as essentially the same thing. On Earth, the more massive an object, the greater is its weight. This thinking was compounded by the fact that almost all of our instruments that measure mass rely on weight. Yet in strict terms, mass is a measure of the amount of matter an object contains; weight is a measure of the force an object exerts at rest in a gravitational field (like Earth's). A set of scales is actually measuring the force applied by an object under gravity.

Unfortunately, once you leave Earth, the concept of weight becomes rather tenuous because it now depends on where in the universe you are. Go to the Moon and weight will be one-sixth of what it was on Earth, yet the mass of everything remains the same. If an object of 1,000kg mass strikes you horizontally at 10m/sec, it will do just as much damage on the Moon as it would on Earth.

RIGHT A page from the flight evaluation report for SA-506, Apollo 11's launch vehicle. The values for the specific impulse of the five F-1 engines are highlighted. *(NASA)*

Table 5-1. S-IC Engine Performance Deviations

PARAMETER	ENGINE	PREDICTED	RECONSTRUCTION ANALYSIS	DEVIATION PERCENT	AVERAGE DEVIATION PERCENT
Thrust 10^3 N (10^3 lbf)	1	6727 (1512)	6740 (1515)	0.198	
	2	6695 (1505)	6674 (1500)	-0.332	
	3	6717 (1510)	6725 (1512)	0.132	-0.027
	4	6748 (1517)	6783 (1525)	0.527	
	5	6717 (1510)	6674 (1500)	-0.662	
Specific Impulse N-s/kg (lbf-s/lbm)	1	2598 (264.9)	2599 (265.0)	0.038	
	2	2599 (265.0)	2598 (264.9)	-0.038	
	3	2596 (264.7)	2596 (264.7)	0	-0.015
	4	2594 (264.5)	2595 (264.6)	0.038	
	5	2587 (263.8)	2584 (263.5)	-0.114	
Total Flowrate kg/s (lbm/s)	1	2589 (5708)	2594 (5718)	0.175	
	2	2576 (5679)	2569 (5664)	-0.264	
	3	2587 (5703)	2590 (5711)	0.140	-0.025
	4	2602 (5737)	2613 (5761)	0.418	
	5	2597 (5725)	2582 (5691)	-0.594	
Mixture Ratio LOX/Fuel	1	2.258	2.255	-0.133	
	2	2.244	2.241	-0.134	
	3	2.262	2.259	-0.133	-0.133
	4	2.254	2.251	-0.133	
	5	2.282	2.279	-0.131	

NOTE: Performance levels were reduced to standard sea level and pump inlet conditions at 35 to 38 seconds.

For the moment, let's deal only with mass in a space environment – no Earth, no Moon. By burning a quantity of propellant in a rocket engine, we turn that propellant into a gas that accelerates and then leaves the nozzle at high speed. In so doing, the propellant has gained momentum. Since momentum is mass times velocity, then the momentum gained by a kilogram of exhaust gas depends solely on its exit velocity. The faster it leaves, the greater its momentum. In this scheme, the specific impulse of an engine is given by the momentum gained for each unit of mass consumed:

Specific impulse = momentum / mass

or by expressing momentum as mass times velocity:

Specific impulse = mass × velocity / mass

The two occurrences of mass cancel out, leaving us with this:

Specific impulse = velocity

Using the SI system, this is stated in metres per second and it means that the efficiency of a rocket engine is simply the speed at which its exhaust exits the nozzle.

Newton's Third Law tells us that for every action, there is an equal and opposite reaction. This means that the momentum gained by the exhaust in one direction must equal the momentum gained by the vehicle in the opposite direction. Therefore, the faster the exhaust gases leave, the greater the momentum that is gained by the vehicle.

Now all this is very neat and tidy, but it doesn't represent the way many rocket engineers think of specific impulse. Traditionally, they have approached it from the point of view of weight, not mass. In this scheme, specific impulse is the ratio of thrust to the weight of propellant flowing through the engine each second and it is normally stated in pounds.

Specific impulse = thrust / weight per second

Or put another way:

Specific impulse = pounds-force / pounds-weight per second

As an example, the thrust of an F-1 engine is 1,500,000 pounds (force) which comes from 5,660 pounds (weight) of propellant passing through each second. The ratio of these two is 265. Notice that there are 'pounds' on both sides of this ratio. On Earth the pound-weight is equivalent to the pound-force, so they cancel out to leave seconds as the unit for this number.

Let's try the same thing using SI units. The F-1 thrust is 6,700,000 newtons and flowrate is 2,567 kilograms (mass) each second. This yields 2,610 and the units would be N/kg/s which can also be written as N-s/kg. But that is mass, not weight. If we want to express that as weight (force), we need to multiply the mass by Earth's gravitational acceleration (a value of 9.8m/sec^2) according to the very famous equation, $F = ma$.

Specific impulse = 6,700,000 / 2,567 × 9.8

The ratio this way is, as should be expected, 266, the same as the above answer except for rounding errors.

But let's return to that number we got above, 2,610N-s/kg, and think of this in terms of momentum. To restate, specific impulse is sometimes thought of as momentum (mass times velocity) gained for unit mass expelled. We don't know the velocity but we do know how to get it because we know the acceleration; it is force divided by mass. For the F-1, we know the thrust (6.7MN) and the mass per second (2,567kg). From this, we get the acceleration of the gases over that time. That's 2,610m/sec, a number which we met before as N-s/kg.

Thus, by applying Earth's gravitational acceleration, we've been able to get to the same place of specific impulse expressed in seconds. Since both use the second as a unit of time, this gives us a conduit between the worlds of imperial and SI units.

To sum up, specific impulse is conventionally stated in seconds, both in SI and imperial systems. It is directly related to the exhaust velocity of the engine such that, if you know the I_{sp} of an engine, you can work out how fast the gases leave by multiplying by 9.8, Earth's gravitational acceleration.

Looked at in terms of specific impulse, the J-2 engine was far more efficient than the F-1. Its figure was about 424 seconds (versus 265 seconds for the F-1), and this translated into an exit velocity of 4,155m/sec.

RIGHT Abe Silverstein (1908–2001), director of the then NASA Lewis Research Center, was pivotal to hydrogen/oxygen propulsion being chosen for the Saturn V's upper stages. *(NASA)*

BELOW An S-IV stage, with its six RL-10 engines, being prepared for shipping to Cape Canaveral. This stage flew as part of the SA-7 launch vehicle on 18 September 1964. *(NASA-MSFC)*

possible roles. Already enthusiastic about LOX/LH$_2$ propulsion, Silverstein was highly influential in recommending these cryogenic propellants for the Saturn's upper stages.

In addition to early thoughts of using a Centaur in a one- or two-engine configuration as one of the upper stages for the Saturn I, the Marshall Space Flight Center envisaged a new rocket stage, the S-IV. This would cluster six RL-10s to form a large and powerful upper stage for the distinctive multi-tanked S-I stage.

Towards the J-2

With a great deal of development experience in LH$_2$ already available, in late 1959 NASA began to explore the possibilities for a much larger LH$_2$ engine. The RL-10 was a relatively low-power unit and if manned space travel was to develop beyond initial experiments, very heavy payloads were in the offing. It was clear that something stronger would be required.

Initially, NASA envisaged an engine ten times more powerful than the then RL-10 with a thrust of 665kN (150,000lb-f) but like the F-1, ambitions grew still further and the target was set at 890kN (200,000lb-f). This engine would be designated J-2 and in September 1960, Rocketdyne was hired to develop it. The contract was notable because for the first time, maximum safety for manned flight was a stated clause in the development of a powerful rocket engine. This was avowedly an engine to carry humans.

J-2 development

Rocketdyne was already experienced in the development of large rocket engines, having produced the kerosene-fuelled H-1 for the eight-engined first stage of the Saturn I. The thrust of that engine was similar to the J-2. The company was also tackling and taming the enormous power of the F-1. But as a fully cryogenic engine that would operate in a vacuum, the J-2 introduced new challenges. For example, a large vacuum chamber had to be built in order to allow components to be tested in their intended environment.

Nevertheless, with Rocketdyne's background in the field, the development of the J-2 advanced at a rapid pace. Within two months, an experimental injector had been produced. By six months, firings were being carried out with an uncooled chamber. Within a year, a working engine was running for over four minutes.

Major problems had to be dealt with along the way. For example, Rocketdyne began with a solid, copper-faced injector similar to that used on their conventional engines like the

H-1 and F-1, only to discover that the thermal conditions within a hydrogen-burning engine are substantially different to one burning RP-1. The spectacular green flames emanating from the engine were a sure sign that the flame front was burning through the injector.

Pratt & Whitney had dealt with this issue in the RL-10 by constructing the injector using porous metal and passing gaseous hydrogen through it to keep it cool. Reluctant at first to accept ideas from outside the company, Rocketdyne had to be forced by NASA to swallow their institutional pride and embrace the concept of the porous injector.

As a fully cryogenic engine, the J-2 presented a range of difficult thermal problems. At LOX temperatures (-183°C) an insulating blanket of frost tended to form around ducts, lines and manifolds. But LH_2 was so cold (-253°C) that air itself would liquefy and pour off the chilled components, exposing them to more air and bringing more heat to the component. It was like a conveyor belt of unwanted warmth. Rocketdyne therefore insulated all LH_2-carrying ducts with vacuum jackets.

However, this solution became tricky where two 20cm diameter ducts brought LOX and LH_2 to the engine. The propellant ducts had to accommodate the bending and expansion/contraction motions created as an engine gimballed through an angle of over 10° to steer the rocket. The solution was to use a double set of metal bellows with a vacuum between as a jacket. A jointed framework around the duct provided additional support.

Like other engines, regenerative cooling was adopted by passing fuel through pipes that formed the chamber walls. A major difference with this large hydrogen engine was that the liquid fuel became a gas part way through its passage, and did so in large quantities. Engineers used computer design, at that time a very novel technique, to come to terms with the complex fluid dynamics involved when designing the pipework and its cross-sectional profile.

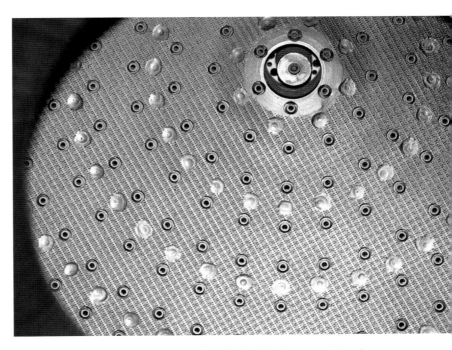

ABOVE The injector of an RL-10 engine. Most of the face consists of a porous metal which permits a small fraction of the gaseous hydrogen fuel to pass. Each LOX injector is surrounded by an annular gap through which passed most of the fuel. *(Photo courtesy of Justin LaFountain)*

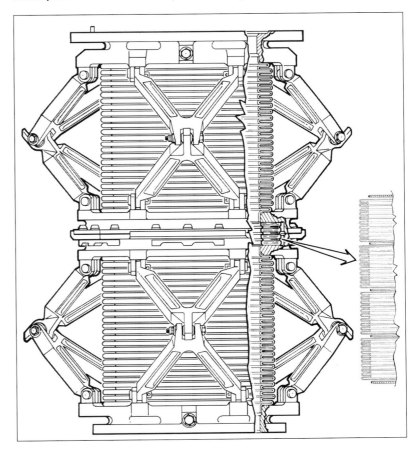

RIGHT Drawing of the LOX duct leading to the J-2 engine. The duct was formed from double walled bellows with a jointed framework to give them stability as the engine swivelled. *(Rocketdyne/NASA/Mike Jetzer)*

LEFT The thrust chamber of J-2 engine number J-2111 stored outdoors at MSFC, as shown by the corrosion appearing on its surfaces. *(Mike Jetzer-heroicrelics.org)*

Engine description

At first glance, the J-2 is like a smaller F-1. It was built around a large bell-shaped thrust chamber, 272cm tall, made of brazed piping around which pumps, a gas generator and various ducts were mounted. There the comparisons cease, because the exotic nature of the fuel mandated profound differences.

Regenerative cooling of the engine was achieved by passing the supercold fuel through the walls of the thrust chamber on its way to being consumed. Consequently, the chamber was constructed from a network of stainless steel tubing with walls only 0.3mm thick for efficient heat transfer. The top of the thrust chamber comprised a 47cm diameter

BELOW Drawing showing two sides of the J-2 engine. The colours indicate the flow paths for fuel, LOX, turbine exhaust and the contents of the start tank which helped bring the engine up to speed. *(NASA/Mike Jetzer/Woods)*

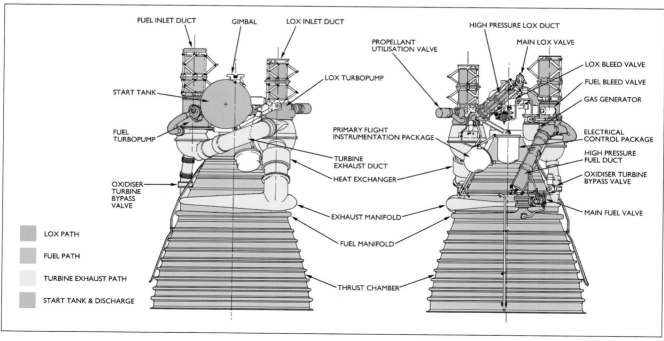

cylindrical combustion chamber. After narrowing at the throat to 37cm, the chamber opened out to a nozzle which had an internal diameter of 196cm at the exit plane. The resultant large expansion ratio of 27.5 reveals that this engine was intended for operation in a vacuum.

A notable difference from the F-1 is that fuel was not introduced at the top of the chamber piping but at a manifold part way down. It therefore made a half pass to the bottom of the nozzle via 180 'down' tubes and a full pass all the way to the top via 360 'up' tubes before entering the chamber and being burned. The doubling of the tubes for the up pass helped to accommodate the change in phase from liquid to gaseous hydrogen (GH_2).

The top of the thrust chamber was capped with an injector assembly that incorporated 614 coaxial injectors surrounded by an injector face of sintered porous metal. Above that was the LOX dome which ushered the oxidiser towards the injector and transmitted the engine's thrust from the injector to a gimbal mount and on to the structure of the rocket. The gimbal mount was a simple universal joint that facilitated steering the engine.

Injector

The injector used a method of merging the fuel and LOX that was very different from the angled orifices used in the F-1. Instead, the J-2 had *coaxial injectors*. The main part of the injector was fabricated from a single piece of metal and included 614 tubes through which LOX was passed under pressure. These LOX tubes were surrounded

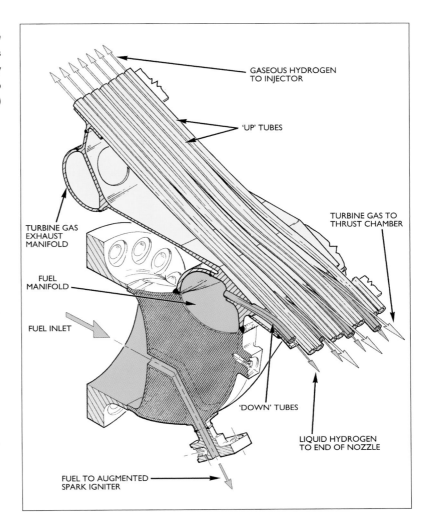

RIGHT Drawing to show the routing of fuel in the thrust chamber pipework. Liquid hydrogen goes down and gaseous hydrogen comes up. The entry of turbine exhaust gases into the chamber is also shown. *(Rocketdyne/NASA/Mike Jetzer/Woods)*

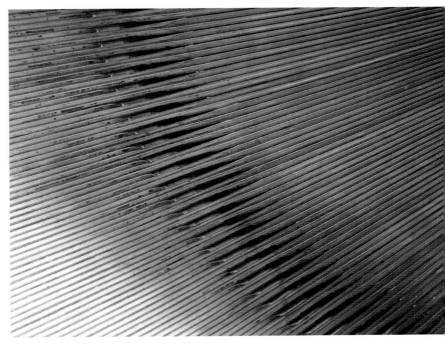

RIGHT The interior of the J-2 thrust chamber showing where fuel enters the pipework within every third tube and the gaps where turbine exhaust enters the nozzle to complete combustion. *(Mike Jetzer-heroicrelics.org)*

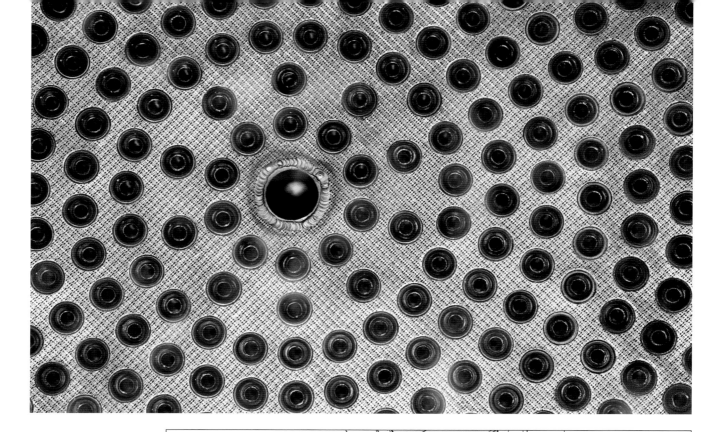

ABOVE Close-up of a J-2 injector. LOX enters at 614 posts, each surrounded by annular gaps where gaseous hydrogen enters, atomising the LOX stream. The rest of the face is porous, allowing about 3.5% of the hydrogen to pass through. *(Mike Jetzer-heroicrelics.org)*

RIGHT Detail of the first set of static vanes in the LOX turbine. These vanes serve to direct the gas flow on to the concave side of the rotating vanes. *(Rocketdyne/NASA/Mike Jetzer/Woods)*

by an injector face of porous metal. A circular gap formed an orifice around each LOX tube through which GH_2 was pumped to form an annular flow around the LOX flow. The LOX was atomised as these two flows met, thereby ensuring the propellants were efficiently mixed for combustion.

The tips of the LOX tubes were slightly recessed below the injector face to improve combustion stability. Further stability was provided by arranging an exit velocity for the GH_2 that was faster than that for the LOX. As in the case of the RL-10 engine, about 3.5% of the GH_2 was allowed to leak through the porous injector to keep it cool.

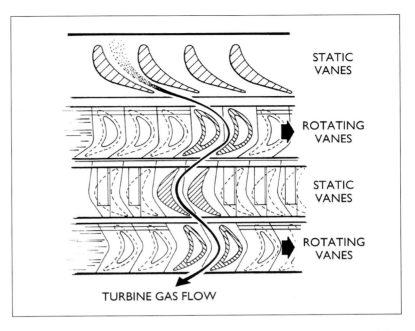

ABOVE Diagram of the gas flow through the static and rotating vanes of the LOX turbopump's turbine. *(Rocketdyne/NASA/Mike Jetzer/Woods)*

BELOW J-2 engine inspection area at the Rocketdyne plant, Canoga Park, California. *(NASA/MSFC)*

73

J-2: EFFICIENCY IN AN ENGINE

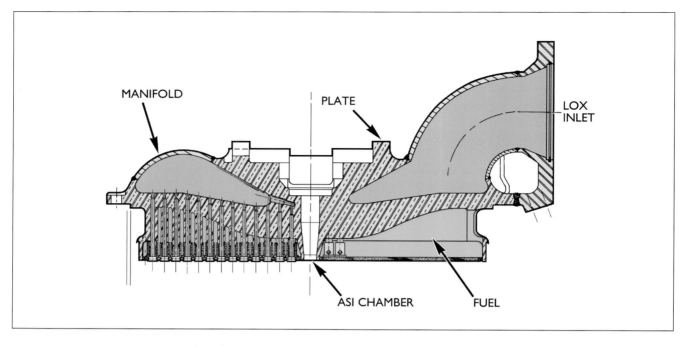

ABOVE Cutaway drawing of the J-2 injector. *(Rocketdyne/NASA/Mike Jetzer/Woods)*

BELOW Detail cutaway drawing of a J-2 injector post. LOX enters through a post just below the injector surface. A collar (yellow) around the end of the post has holes to permit gaseous hydrogen to pass and it exits in an annular ring around the LOX, atomising it prior to combustion. *(Rocketdyne/NASA/Mike Jetzer/Woods)*

BELOW Close-up photo showing a sectioned J-2 injector. This shows the holes in the collars where fuel passes on its way to the thrust chamber. *(Mike Jetzer-heroicrelics.org)*

BOTTOM Close-up of the face of a J-2 injector. LOX passes from the central pipe of each injector. Gaseous hydrogen exits through the gap around each pipe. About 3.5% of the hydrogen passed through the porous injector face in order to keep it cool. *(Mike Jetzer-heroicrelics.org)*

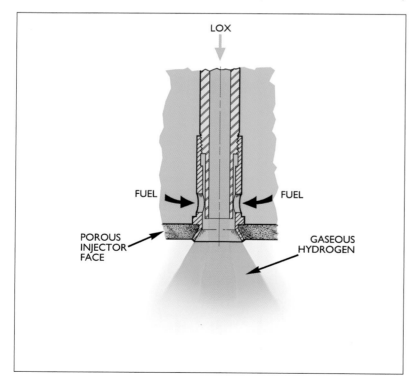

Turbopumps

Whereas the F-1 combined both fuel and LOX pumps into a single unit driven by one turbine, the arrangement for the J-2 was very different. Because hydrogen is much less dense than oxygen, it called for an alternative means of raising its pressure. This translated into a much higher turbine speed and hence a separate turbopump.

A gearing system could have driven both pumps from a single turbine but this would have added mechanical complexity. Also the extreme cryogenics involved would have presented a lubrication problem.

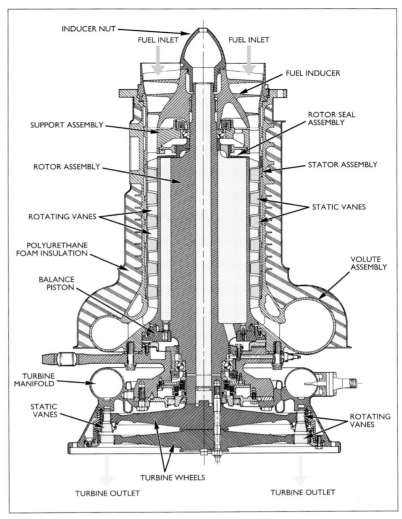

RIGHT Cutaway drawing of the J-2's fuel turbopump. LH$_2$ entered from the top, passing through an inducer, seven stages of rotating vanes, then a volute, prior to being passed to the thrust chamber. Two turbine wheels powered the pump. *(Rocketdyne/NASA/Mike Jetzer/Woods)*

BELOW A J-2 fuel turbopump. Fuel entered from the top and exited from the flanged outlet on the left. *(Mike Jetzer-heroicrelics.org)*

BELOW RIGHT A 3D drawing of the major components of the J-2 fuel turbopump. This shows the arrangement of rotating and static vanes. *(Rocketdyne/NASA/Mike Jetzer/Woods)*

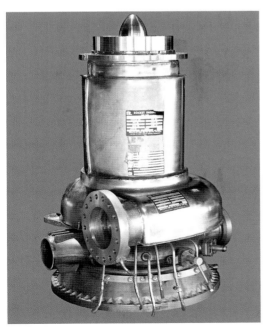

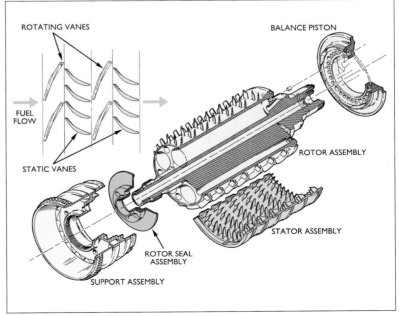

J-2: EFFICIENCY IN AN ENGINE

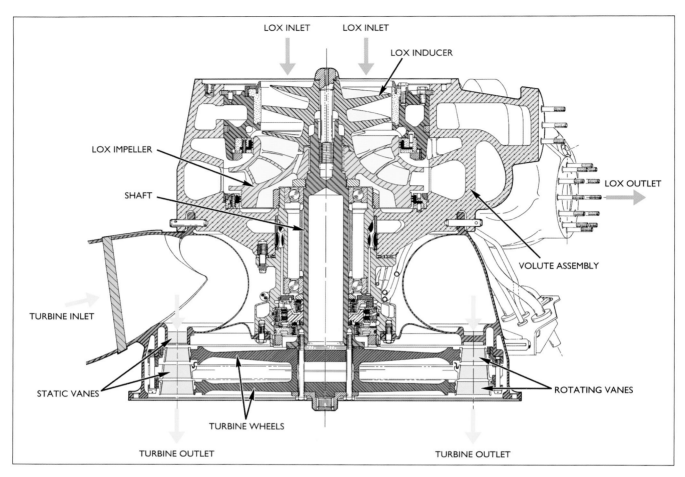

ABOVE Cutaway drawing of the J-2's LOX turbopump. LOX entered from the top, through an inducer and impeller, before entering a volute from which it exited at high pressure for the LOX dome. Two turbine wheels powered the pump. *(Rocketdyne/NASA/Mike Jetzer/Woods)*

The two turbopumps were mounted on opposite sides of the thrust chamber. Like the F-1's pumps, they were powered by a gas generator that burned a small amount of the available propellants. The resulting exhaust gas was first fed through the turbine that drove the fuel pump, and then onto the turbine of the LOX pump. Between them, the two pumps drove approximately a quarter of a metric ton of propellant into the engine per second.

The fuel pump was of the axial type; i.e. it used multiple disks of fan vanes to push the LH_2 along a cylinder concentric with the axis of the pump, similar to the compressor of a jet engine. This rotor was directly driven by a shaft from a two-stage turbine. Gas generator exhaust passed through two turbine wheels

LEFT A J-2 LOX turbopump. The inlet at the top has a cover over it. The LOX outlet is to the right and the turbine inlet is on the left.
(Mike Jetzer-heroicrelics.org)

to spin them, and hence the pump rotor, at 27,000rpm. A set of static vanes located between the turbine wheels straightened the gas flow. The pump operated at 5.8MW (7,800 brake-horsepower) to raise the LH_2 pressure to 8,446kPa (1,225psi).

After leaving the fuel pump's turbine, the exhaust from the gas generator went on to power the LOX turbopump on the opposite side of the engine. The turbine for the LOX pump was also a two-stage arrangement with a stator section between. It rotated at the much slower speed of 8,600rpm to produce 1.6MW (2,200 brake-horsepower). The turbine directly drove the pump rotor which was a centrifugal arrangement of inducer and impeller, much like that in the F-1, to raise the LOX pressure to 7,446kPa (1,080psi).

Both pumps were lubricated by their respective working fluids. This was an important consideration for the LH_2 pump and its extremely low temperature. It also saved weight. A series of seals on the drive shafts prevented them from leaking through to their respective turbines. Dynamic seals were often used whereby, as the pump ran, extra vanes on the shaft created sufficient pressure to counter any tendency for fluids to move in unwanted directions.

Gas generator

Turbine gas came from a generator attached directly to the housing for the fuel turbine. Inside this unit, two spark plugs ensured the propellants were ignited, whereupon the fuel-rich exhaust gas passed through the turbine sections of both turbopumps. Having fulfilled its chief role of powering the pumps, the gas then passed through a heat exchanger where its heat converted LOX to gaseous oxygen in order to pressurise the LOX tank.

The final destination for the turbine exhaust was to be fed into a wraparound manifold in a manner similar to what occurred in the F-1 and then fed into the thrust chamber part way down the nozzle, adjacent to where the fuel pipes arrived at the thrust chamber wall. However, its fate here was not to protect a surface. Instead the fuel-rich gas completed its combustion and in so doing, further improved the engine's efficiency.

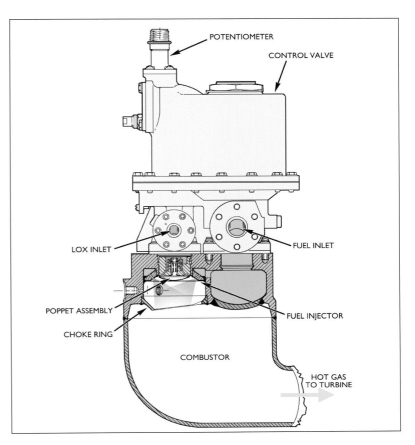

ABOVE Cutaway drawing of the J-2's gas generator. LOX and fuel entered the combustor from the top, the exhaust exiting bottom right for the fuel turbopump. *(Rocketdyne/NASA/Mike Jetzer/Woods)*

BELOW Cutaway graphic of the J-2's heat exchanger. This used the heat of the turbine exhaust to turn LOX into gaseous oxygen in order to pressurise the LOX tank. *(Rocketdyne/NASA/Mike Jetzer/Woods)*

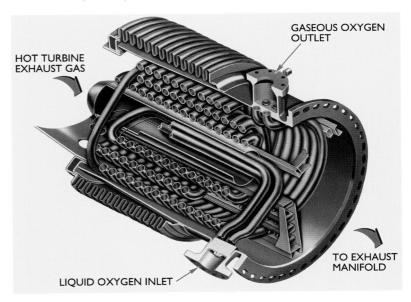

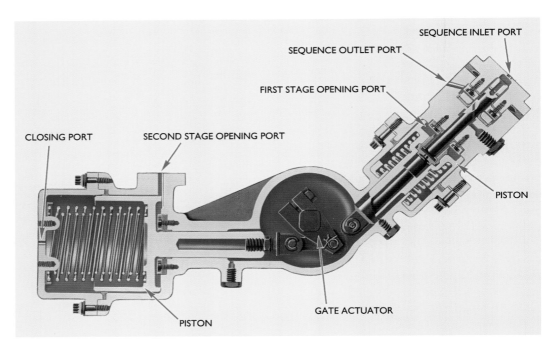

RIGHT Cutaway graphic of the J-2's LOX valve actuator. Opening or closing pressure at the left moved a piston which turned a crank. The crank then operated a sequence valve to help coordinate the start of the engine. (Rocketdyne/NASA/Mike Jetzer/Woods)

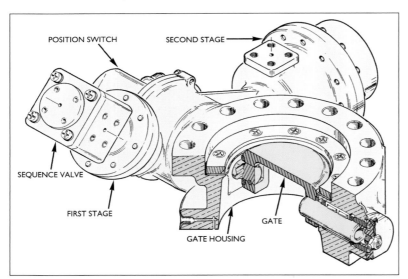

BELOW Isometric cutaway drawing of the main LOX valve showing the valve gate. The actuator (illustrated above) is on the far side of the valve itself. (Rocketdyne/NASA/Mike Jetzer/Woods)

Main valves

Two insulated ducts routed the propellants from their respective pumps to the thrust chamber. One duct fed the LOX dome at the top of the chamber. The other duct took LH_2 on a longer route to a manifold that wrapped around the chamber halfway between the throat and exit plane. Here it was fed into the regenerative cooling system. Within those ducts were the main engine valves; a LOX valve and a fuel valve.

Both were of the butterfly type; i.e. the central shutoff mechanism was a disk that sat in the flow. By rotating 90° the disk could open or shut the valve. These valves were normally shut under the action of a spring and opened by the application of pneumatic pressure.

In a fashion similar to the F-1's LOX valve, each main valve included a sequence valve mechanism. This sensed the position of the butterfly in order to ensure that another event could only occur at a defined stage in the start sequence, when a valve had opened to a particular extent.

In addition to these basic elements of the liquid rocket engine, the J-2 had a number of subsidiary subsystems to help it function.

Start tank

Apollo was all about going to the Moon and that meant leaving the gravitational safety of the home planet. The manner in which a mission was conducted required that the spacecraft park in low Earth orbit in order to have its systems checked out. Only then would the crew set out for the Moon using a manoeuvre known as *translunar injection*.

This manoeuvre was carried out by the Saturn's S-IVB third stage and it required that the stage's single J-2 engine be restarted in space, having already been used to place the stack in Earth orbit. Indeed, all the J-2 engines that were supplied to NASA had this capability but it was not enabled on the five engines that powered the S-II stage because they were fired only once in a mission. As an engineering test, the J-2 engine on Apollo 9's S-IVB was

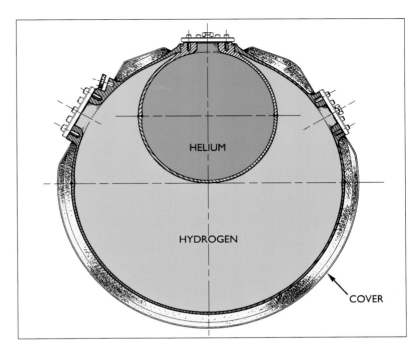

RIGHT **Cutaway drawing of the J-2 start tank. This was filled with high-pressure hydrogen which, when discharged through the turbines, began to power up the engine. A helium tank was built within to provide pneumatic pressure to operate the engine valves.**
(Rocketdyne/NASA/Mike Jetzer/Woods)

restarted a second time. Otherwise, only one restart was required for a normal lunar mission.

The trick with starting and restarting a J-2 was to get its turbines spinning. Once the turbine wheels were rotating fast enough, the power they supplied would maintain the necessary pressures to keep the engine running until it was commanded to shut down.

Energy to spin up the turbines from a standing start came from a tank of high-pressure hydrogen gas that was filled prior to launch. Appropriately, this was known as the *start tank*. A duct leading from the tank included a valve that would open to discharge the contents of the tank upon command, so this was known as the *start tank discharge valve* (STDV).

When this valve opened, GH_2 from the start tank entered the pipework that led to the turbines. It passed through each turbine in turn to get them spinning. From that point on, there was sufficient power available from the turbopumps to force propellants into the gas generator. Ignited by spark plugs, the resultant combustion gas then took over the task of powering the turbines, bringing the engine up to speed.

That got the engine started but it didn't in itself permit it to start again. To do so, the start tank had to be refilled with GH_2. Fortunately, while the engine was operating it generated a ready supply of high-pressure hydrogen in the pipework of the thrust chamber. Two feeds were therefore installed from the thrust chamber to the start tank. One feed tapped high-pressure liquid hydrogen from the point where it entered the thrust chamber. The other feed tapped high-pressure gaseous hydrogen from a point after the hydrogen had been vaporised by passing through the jacket of the chamber.

These feeds of liquid and gas were permitted to flow into the start tank for a period of one minute while the engine operated. The mix was

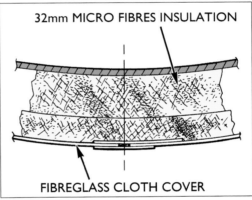

LEFT **Cross section detail of the start tank wall.**
(Rocketdyne/NASA/Mike Jetzer/Woods)

controlled to keep start tank temperatures and pressures within limits when restarting the engine.

Tank within a tank

The hydrogen-filled start tank actually contained another tank. This internal tank stored helium gas to provide the pneumatic power required to operate the engine's valves. By placing one tank within the other, the layout of the engine was simplified. Moreover, helium's tendency to leak was limited by keeping the tank under pressure and at low temperature.

Oxidiser turbine bypass

Since the two turbopumps were driven from a single supply of turbine gas, there had to be a means of balancing their power. Engineers therefore included a duct that allowed a fraction of the turbine gas to bypass the oxidiser turbine in a controlled fashion. A valve within the duct could adjust the effectiveness of this bypass.

During start-up, the valve was kept open to allow a large fraction of the turbine gas to bypass the oxidiser turbine. This restricted the acceleration of the oxidiser pump which meant that initially, mostly fuel was being introduced into the chamber.

In the final moments of engine start, the bypass valve closed, blocking the bypass duct and forcing almost all the turbine gas through the oxidiser turbine to bring the oxidiser pump up to speed. But a hole in the valve gate permitted a small flow of turbine gas to continue through the bypass duct. By setting the size of this hole during engine tests, engineers could calibrate the power balance of the two pumps for a particular engine.

Mixture ratio control valve

Unlike the main engines on the Space Shuttle, which could be throttled, the engines on the Saturn V were classed as fixed-thrust engines. This was especially true for the F-1, but for the J-2 there was a means of making small adjustments to its thrust and efficiency while it was in operation. Engineers utilised this capability to gain more performance from the S-II and S-IVB stages.

In burning two propellants in a rocket engine, engineers must choose the most appropriate *mixture ratio* (MR): how much oxidiser is burned with how much fuel. At first look, it might appear that the best MR is that which burns the fuel most completely. This is known as the *stoichiometric ratio* and for a hydrogen-oxygen combination the most complete combustion occurs at a mixture ratio of 8:1 because each oxygen atom (atomic weight = 16) requires two hydrogen atoms (atomic weight = 1) in order to make a molecule of water. Burning 16kg of LOX with 2kg of LH_2 will produce 18kg of H_2O and extract the maximum possible chemical energy.

However, other factors influence the choice of ratio. These include combustion temperature and the exhaust velocity (and hence specific impulse). For the J-2, a ratio of 5:1 was chosen; 5kg of oxygen and 1kg of hydrogen with a quantity of unburned hydrogen leaving with the exhaust. This yielded a thrust of about 890kN (200,000lb-f) and a specific impulse of 427. Hence the exhaust velocity was 4,185m/sec.

By adding a small duct between the oxidiser pump's outlet and its inlet, and feeding a small amount of high-pressure LOX back to the inlet, engineers affected the ability of the pump to do its job well. Further, a valve in this duct permitted control of this effect. This allowed a degree of adjustment of the mixture ratio and therefore its thrust and efficiency.

Early engines had a three-position valve that gave MR values of 4.5:1, 5.0:1 and 5.5:1. The engines on the S-IIs from Apollo 14 onwards were simplified to two positions; a low MR setting (4.8:1) when it was open, and a high setting (5.5:1) when it was closed. The J-2 engines on the S-IVBs were only ever burned at 4.5 or 5.0 MR.

Changing the MR had a number of effects. When an early engine was set to a high MR of 5.5:1, its thrust rose to 1MN (225,000lb-f) because there was more oxygen to react with the hydrogen. This created more energy but also reduced the specific impulse to 424 seconds. This was because water molecules are relatively heavy and require more energy to accelerate them. Thus the exhaust velocity was low and so was the efficiency.

Moving to a low MR of 4.5:1 brought the thrust down to only 0.79MN but because there was more unburnt hydrogen in the exhaust and because hydrogen is light and can therefore be accelerated to faster exhaust velocities, the

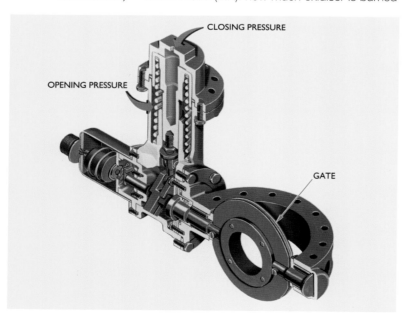

BELOW Cutaway graphic of the J-2's LOX turbine bypass valve. An open gate partially bypassed the LOX turbine, slowing its speed on start-up. When closed, the size of the gate's orifice allowed the full speed of the turbine to be set to balance the fuel turbopump. *(Rocketdyne/NASA/Mike Jetzer/Woods)*

specific impulse was increased to 430 seconds, thereby improving the efficiency of the engine.

The chief reason for changing the MR, however, was to allow control over the rate at which the two propellants were consumed. Then, by making the change at the right time, their consumption could be balanced in order to, as far as possible, ensure both propellants were depleted at the same time. The strategies behind the use of this capability differed between the S-II and the S-IVB stages and are discussed within their relevant chapters.

Propellant bleed

Use of cryogenic propellants, particularly hydrogen, required that the propellant lines and engine propellant ducts be cooled prior to engine start. This was to avoid the premature formation of gas and prevent cavitation within the pumps.

A recirculating flow of propellant was set up in the tanks and through the feed ducts to maintain cold conditions throughout. This flow was extended into the engine by opening bleed valves to ensure a return flow through the pumps. The valves were closed during engine operation.

Augmented spark igniter

Liquid hydrogen and oxygen, although an extremely flammable combination, are not hypergolic. They require an ignition source to initiate combustion. In the gas generator, ignition was achieved by a pair of spark plugs. For the thrust chamber, two spark plugs were also used but they ignited a dedicated feed of propellant.

This was the *augmented spark igniter* (ASI) and it was mounted at the centre of the injector. After the spark plugs were switched off, the flame produced by the ASI was maintained to ensure that combustion was sustained across the injector's face.

Starting the J-2

The engineering term for when an engine was running at its specified power was *mainstage*, and bringing a J-2 engine to this condition was every bit as complicated and orchestrated as for the F-1. Further, one detail of the transition to mainstage was altered depending on whether the engine was on the S-II or the S-IVB, and in the latter case, whether it was being ignited for its first or second burn.

The show began with a signal from the Saturn V's instrument unit commanding the engine to start. This triggered the warm-up act, mainly by the action of the *mainstage control valve*, an electrically operated mechanism that correctly sequenced the pneumatic operations of the various players that would follow.

Two spark plugs in the gas generator and two in the augmented spark igniter at the centre of the injector were energised. These would continue to deliver sparks throughout the start-up until de-energised by a timer.

Simultaneously, helium drawn from the tank-within-a-tank at a regulated pressure was allowed into the pneumatic control system to operate the engine's valves. As a precaution, an accumulator was filled with helium to continue to supply pressure in the event of a loss of pressure from the source tank.

This helium pressure was used to close bleed valves to stop the slow recirculation of propellant through the stationary pumps. Up to now, this flow had been keeping the pumps and associated ducts chilled.

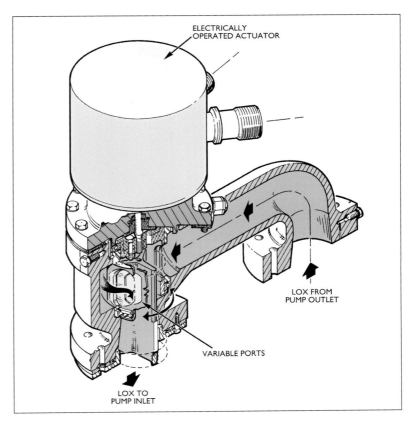

ABOVE Isometric cutaway drawing of the mixture ratio control valve, also known as the propellant utilisation valve. The unit allowed a controlled flow of LOX from its pump's outlet to the inlet, thereby adjusting the pressure being delivered to the thrust chamber. *(Rocketdyne/NASA/Mike Jetzer/Woods)*

ABOVE A J-2 engine being tested. Note the relatively colourless nature of the flame. *(NASA-MSFC)*

The pressure opened the LOX turbine bypass valve so that when the pumps did begin to spin, the LOX pump's speed would be held back, injecting mostly fuel into the combustion chamber. With this valve open, some of the turbine gas would, at least initially, bypass the LOX turbine. A feed of helium was also used to purge the LOX dome and the oxidiser passages of the gas generator.

The start command triggered another precursor to the main show, a timer that delayed the operation of the start tank's discharge valve (STDV). This timer ran for either 1, 3 or 8 seconds depending on circumstance to allow enough time for the supercold fuel to circulate through the myriad of pipes in the walls of the thrust chamber, conditioning them for the flow to come when the pumps were spun up.

For an S-II engine, the delay was only one second. The five engines within their interstage section were already quite chilled by this time and, being in a cluster, they tended to keep each other cold. The single S-IVB engine had a 3-second delay for its first start, as it was isolated in its interstage and needed a little longer for the chill-down. For the second S-IVB start, an 8-second delay was required because its pipework was likely to have been warmed by the Sun by being exposed for nearly three hours in orbit.

To establish a flow of fuel through the thrust chamber's walls, the main fuel valve began to open. This also sent fuel to the ASI. Another valve was opened to feed LOX to the ASI, whereupon the spark plugs ignited the two propellants to create a flame at the centre of the injector face.

At last, the initial preparations were in place and the main show could begin. The ASI flame had been successfully lit and both propellants were flowing slowly past stationary pump rotors. As the main fuel valve reached its 90% position, it operated a sequence valve built into its mechanism to allow helium pressure through to another valve that would control the operation of the STDV. At the point that the STDV timer ran out, a test was made to see that an *enable* signal was present. If the test was positive, the STDV was commanded to open.

Now the action began to pick up as the start tank discharged its hydrogen contents into the ducts leading to the turbines. The sudden flow of gas imparted enough rotational energy to the turbines to allow the rest of the start sequence to take place. Thanks to the open bypass duct, the initial speed of the LOX turbine was held back.

Once the engine had detected the successful discharge of the start tank, it opened the main oxidiser valve. Like the fuel valve, this had a sequence valve which was opened when the main mechanism had moved to a predetermined position. As LOX began to flow under the pressure from its tank, helium pressure from the sequence valve opened the propellant valves for the gas generator.

At the same time, the bypass valve for the LOX turbine began to close. However, an orifice in the helium line to the valve restricted the rate at which it closed so that the speed of the LOX turbine would rise in a controlled fashion as the bypass became less effective.

Both LOX and LH_2 entered the gas generator where their combustion was initiated by the two spark plugs within. As the propellants burned, prodigious amounts of exhaust gas raced through ducts to the fuel turbine and, after that, to the LOX turbine.

The increasing speed of the pumps caused the propellant pressures to rise sharply. Both LOX and LH_2 began to spray into the combustion chamber from the 614 coaxial orifices across the injector face, whereupon they were ignited by the flame emanating from the ASI.

As the turbopumps ramped up to their steady-state speed and the J-2's thrust approached its operational level, a *Thrust OK* switch was activated which indicated to the instrument unit, to the astronauts in the spacecraft and to the controllers on the ground that the engine had reached *mainstage*, its designed power output.

As the engine settled into operation, the timer that had been keeping the spark plugs going ran out. They no longer needed to provide an ignition source because the engine would continue to run until it was commanded to stop; an event that would be triggered either by declining propellant levels in the S-II or by the S-IVB having reached a desired velocity.

Engine shutdown

When a command was received from the Saturn's instrument unit for the J-2 to shut down, this initiated a sequence of operations that began by closing the main LOX valve and starting a timer. Other simultaneous operations included a purge by helium gas of the LOX dome and the LOX injector of the gas generator, and opening the duct that bypassed the LOX turbine so as to divert turbine gas away and thereby reduce the effectiveness of the LOX pump.

Next, the LOX feed to the ASI was shut off and the main fuel valve was closed at the same time as a fast shutdown valve was opened. This released helium pressure from most of the engine's valves, allowing them to close under spring pressure. The gas generator stopped producing power for the turbines as its propellant supply was removed.

The expiry of the timer terminated the helium supply. At the same time, the loss of helium pressure allowed the propellant bleed valves to open under spring pressure, thereby re-establishing a flow of propellant back to the tanks to maintain the low temperature of the engine ready for its next burn, if there was one.

J-2X

Throughout the Saturn V programme, the J-2 showed itself to be a hugely successful engine and it laid the technological foundations for the much more powerful and capable hydrogen-burning Space Shuttle Main Engine (SSME). In 2005, NASA initiated a programme to build the restartable J-2X engine for the upper stages of the Space Launch System (SLS).

The J-2X is based on the same operating principle as the J-2, in that it has a separate gas generator to power two separate turbopumps for the hydrogen and oxygen propellants. The design thrust of just over 1.3MN is 28% higher than the J-2 in its most powerful form, with a specific impulse of 448 seconds, an improvement of about 5%. This increase in efficiency is primarily due to a very high expansion ratio of 80:1 provided by a large nozzle extension.

Though the J-2X was intended to be derived from the J-2's design, by the time modern knowledge, practices and standards were applied, it was essentially an all-new engine. It borrowed technology from more recent engines like the SSME, the RS-68 (a powerful and simplified engine used for the Delta IV rocket) and the European Vulcain engine that powers the Ariane V rocket. As of writing, its further development has been shelved as NASA explores other options for the SLS upper stages, including the use of the smaller RL-10 engine, which had itself constantly evolved over decades of use.

BELOW A J-2X engine at the Stennis Space Center (formerly the Mississippi Test Facility) for testing. *(NASA)*

Chapter Five

S-II: The troubled stage

To get to a low Earth orbit, the Apollo spacecraft had to reach a velocity of 7.8km/sec. Planet Earth provided 5% of that by virtue of its rotation, and the S-IC added another 31%. The first burn of the S-IVB third stage completed the task of getting to orbit by contributing 11% of the total velocity required. Therefore, most of the spacecraft's orbital velocity, 53% or 4,115m/sec, came from the efforts of the S-II stage, an engineering masterpiece that nearly brought Apollo to its knees.

OPPOSITE Apollo 10's S-II stage is lifted through the cavernous void of the VAB at the Kennedy Space Center. *(NASA)*

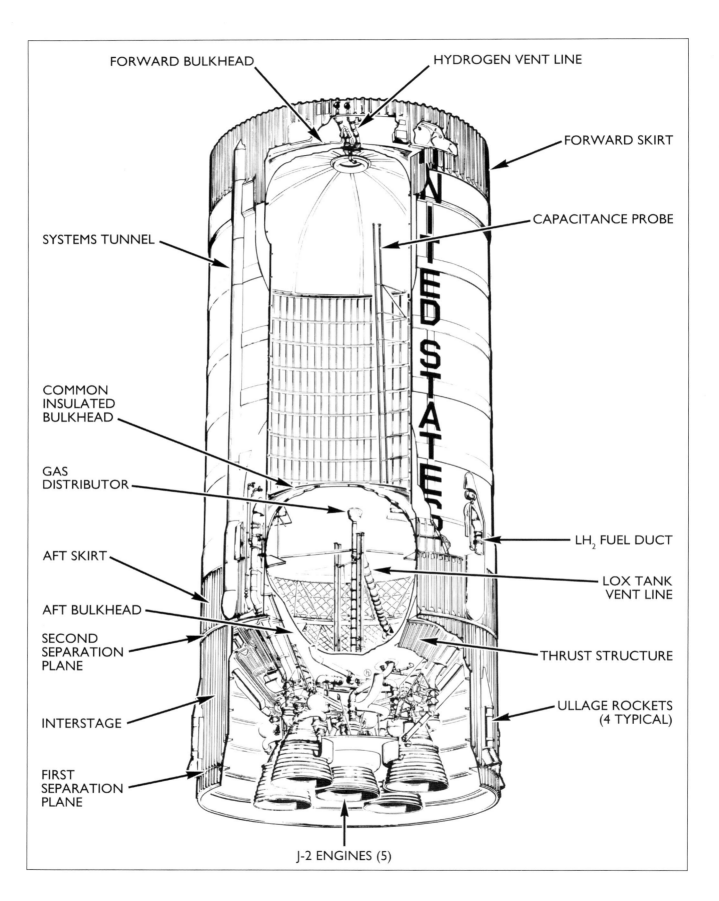

Early ideas about a stage called S-II came in 1959 from a committee, chaired by Abe Silverstein, which was tasked with bringing definition to the Saturn vehicle family. Their C-3 rocket concept included a stage called S-II, and being a persuasive proponent of liquid hydrogen (LH_2) technology, Silverstein thought it should have three or four of the new LH_2/LOX engines he was proposing.

As plans for the Saturn family developed through 1961, the S-II concept became part of a Saturn C-2 configuration. In the process, it was defined to be 22.5m long and 6.5m in diameter to match the S-IB stage. Yet as MSFC tendered for its construction, they were unable to present the prospective contractors with a settled configuration. Even the stage's diameter grew; first to 8.13m and then 10m to match the S-IC. The length was eventually set at 24.9m.

On 11 September 1961, North American Aviation won the contract for the S-II. Soon after, it was also contracted to build the Apollo command and service module (CSM). At the time, many observers doubted the wisdom of saddling the company with so much work. After all, it was now responsible for two very major components of the Apollo system. Nevertheless, a large facility for the production of the S-II was built by the US Navy on North American's behalf at Seal Beach, California.

When elements of the 10m S-II began to be built two years later, the other two stages in the Saturn V stack were much further along in their design and construction. By the mid-1960s, a perfect storm struck the S-II as a confluence of circumstance and accident severely tested North American and those at NASA who oversaw the company's efforts.

Destruction

Inexorably, both the Apollo CSM and the lunar module (the latter being built by Grumman Aircraft Engineering Corporation on Long Island, New York) were getting heavier which caused NASA to demand greater performance from the Saturn V contractors, and a reduction in mass. The S-IVB was in production for the Saturn IB so there was reluctance to change it. The S-IC was also far into its construction cycle and physics dictated that lightening it would have relatively little effect. S-II development had to bear the brunt of NASA's demands.

But the S-II was a very high technology machine using exotic propellants. Although it made use of techniques conceived for the S-IVB it implemented them to a far greater scale, which forced North American to develop new manufacturing approaches on the hoof. As a result of the drive to reduce the stage's mass, the hardware of an S-II stage represented only 7.5% of its loaded weight, the rest being propellant. Compare this to the outwardly similar S-IVB stage at 9.5%.

Very long, difficult welds on huge yet thin skin panels were trying the patience of welders and managers alike. The need to insulate a gigantic tank of supercold liquid hydrogen was causing problems because insulating panels tended to pop off when tested at very low temperatures. At one point, a huge bulkhead that was to separate the LH_2 and LOX compartments came apart under test.

The S-II problems came to a head on 29 September 1965 when a complete stage, specifically built for structural testing and not for flight, ruptured catastrophically and was destroyed. It was being loaded to simulate the stresses that would be endured during the highest acceleration phase of the ascent.

NASA came to the conclusion that the S-II programme was out of control and instituted a 'Tiger Team', a concept borrowed from the Air

OPPOSITE Cutaway drawing of the S-II stage indicating its major components. *(NASA/North American Rockwell/Mike Jetzer/Woods)*

BELOW A still frame from a 16mm film shows the remains of S-II-S/D (structural and dynamic test article) after it was destroyed during a loading test on 29 September 1965. *(NASA)*

LEFT S-II-T (used for live fire testing) being lifted into a test stand at the Mississippi Test Facility. It too would be destroyed when an overpressure ruptured it on 28 May 1966. *(NASA)*

Force, to go into North American Aviation to review the handling of its Apollo contracts and the effectiveness of its management. In the eyes of NASA, progress with the S-II had become the pacing item of the entire Apollo programme.

This close study of North American by some of the leaders of the Apollo programme led to a report named after the head of this team, General Sam Phillips. The Phillips Report was a devastating critique of the company and its approach to Apollo, and it caused a shake-up of North American's management. In particular, former Air Force Major General Robert Greer took on responsibility for the S-II, a role for which he would ultimately gain praise – but not yet.

On 28 May 1966, another S-II built specifically for testing was unintentionally overpressurised and destroyed when it subsequently ruptured. Five North American staff were injured in the calamity. The troubles affecting the S-II were beginning to impact the launch schedule for the entire Saturn V. During preparations at KSC for both Apollo 4 and Apollo 6, tests of the launch vehicles had to use a giant spacer in place of the S-II due to the late arrival of the flight stages.

Teams of MSFC engineers began to examine every minute detail of the remaining stages, whether already at the Mississippi Test Facility or part way through manufacture. They found small cracks and leaky seals, plus a plethora of other problems that were suggestive of poor workmanship and quality control by a management that was failing to live up to expectations.

As if things were not already dire enough for North American, on 27 January 1967 the deaths of three astronauts in a lethal oxygen-fed fire in command module No. 012 during ground

LEFT This second stage simulator (or spacer stage) had to be used during checkout of adjacent stage interfaces in the VAB in late 1966 through early 1967. *(NASA)*

LEFT Still frames from a 16mm film show the anticipated rupture of a test tank forward bulkhead during tests of the common bulkhead at Santa Susana, late 1966. *(NASA)*

tests at KSC brought the scrutiny of the press and public onto the company and its NASA overseers. Questions were raised about how one manufacturer had managed to gain two of the most difficult contracts in the Apollo programme. North American eventually responded with a comprehensive rout of its top management.

Around the same time as the spacecraft fire, the first flight-capable example of the Saturn V second stage, S-II-1, arrived at the launch site and was mated with the rest of the first flight-capable Saturn V vehicle, designated AS-501. It was scheduled for launch in August 1967, but this was not to be. Cracks and defects were still being found and the stack was ordered to be taken down for intensive tests and repairs to the S-II.

As is often the case, the three-month delay proved to be beneficial all round because in addition to fixing the problems with the S-II, various repairs and modifications were also made to the other two stages, as well as to ancillary equipment at the launch pad. In the

BELOW S-II-2, the Apollo 6 second stage during stacking in the VAB in mid-1967. An engine flaw and a wiring fault would have profound effects on this stage's performance. Normally the stage was stacked with its interstage but in this case, that had already been installed with the spacer. *(NASA)*

ABOVE The aft of a production S-II shows the crossbeam upon which the centre engine was mounted. *(Mike Jetzer-heroicrelics.org)*

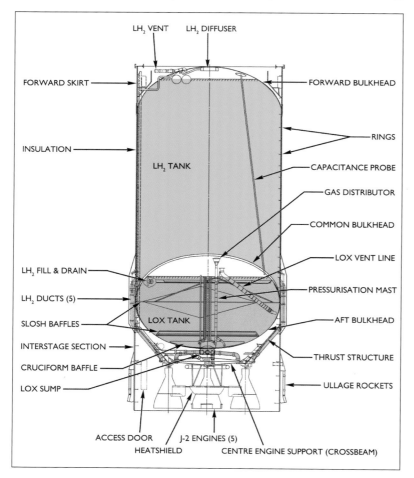

event, AS-501 was launched on 9 November 1967 as Apollo 4. The flight was just about perfect and S-II-1 acquitted itself extremely well.

Across the subsequent twelve flights of the Saturn V, the S-II encountered only two major incidents. When an engine failed on AS-502, launched as Apollo 6 in April 1968, incorrect wiring in the stage caused the wrong engine to be shut down. Worse, the erroneously terminated engine was located on the same side of the cluster as the engine that had failed. It had been thought that the loss of power on two adjacent outboard engines would make the vehicle uncontrollable. Yet the stage managed to compensate for the asymmetric thrust and limped towards orbit using its other three engines, much to the surprise of the Saturn's flight control team.

The second incident struck Apollo 13 on 11 April 1970, when the centre engine experienced pogo vibrations so severe that in responding to ±33.7g acceleration forces, its supporting crossbeam was flexing by "inches". A switch intended to check that the thrust was within spec was tripped and it shut the engine down early. Fortunately, the loss had no operational effect on the mission because the other four engines were able to make use of the remaining propellant to achieve the stage's task.

Engineers reckoned that the cause of this anomaly was reduced pressure in the LOX tank. It made the stage sufficiently unstable to allow cavitation at the engine's LOX inlet. This cavitation, a cycle of forming and collapsing bubbles, amplified the pogo vibrations to the point where the sensor that detected the vibration on the crossbeam was saturated.

Airframe

At its simplest, the structure of the S-II can be thought of as a large tank assembly with a cylindrical forward skirt at one end and an aft skirt at the other to accommodate the bulbous domes at the ends of the tank assembly. The aft skirt included a thrust structure that carried five J-2 engines. Below the aft skirt was a cylindrical interstage ring

LEFT Cross section of the S-II, coloured to identify the propellant routes. *(NASA/Mike Jetzer/Woods)*

that held the S-II away from the S-IC to give clearance for the engines. It was manufactured as part of the stage, but because it was jettisoned early in the S-II's flight it is often regarded as a separate unit.

Along with the interstage, the two skirts were primary structural components and they defined the ends of the stage's cylindrical shape. Constructed using 7075 aluminium alloy, they incorporated longitudinal stringers with a hat-shaped cross-section that were affixed along the outside in order to stiffen the skin. Further stiffening came from a series of internal rings that ran around the inside of all three components. Since the aft skirt and interstage carried a greater load, their skin was 1.8mm thick whereas the forward skirt's skin was just 1mm thick.

The thrust structure came as part of the aft skirt and was constructed from the same 7075 alloy. Its main component was a stiffened cone between two rings. The smaller ring included mounts for the four outer engines in a 5.4m diameter circle. A crossed beam structure straddled this ring between the outboard engine support points to accommodate the fifth engine at the centre. The cone spread the force from the engines to a 10m ring matching the diameter of the stage. Four pairs of longeron members mounted along the cone emanated from each of the outboard engine support points to help distribute the force.

A lightweight but rigid heatshield made of fibreglass honeycomb was mounted between the five J-2 engines. This 5.4m disk inhibited the backflow of hot exhaust gas to the upper parts of the engine and the boxes of equipment mounted around the thrust structure's cone. Cut-outs were made in the disk to accommodate each engine, with flexible silica curtains bridging the gaps to allow for engine gimballing.

Tankage

The tank assembly was 10m wide and comprised two compartments. The lower LOX compartment, ellipsoidal in shape and 6.7m high, was designed to hold 331,000 litres of oxidiser. Above it was the LH_2 compartment which was 17m high and held 1 million litres of fuel. It was fabricated from six cylinders welded together and topped by a domed bulkhead.

Each cylinder was welded from four panels, each milled from a single block of metal to include internal stringers and circumferential rings for strengthening.

Due to its useful property of gaining strength at cryogenic temperatures, 2014-T6 aluminium alloy was chosen for tank fabrication. Unfortunately it is also prone to cracking, and this was a factor in the troubles that beset the S-II during its development.

The domed bulkheads caused particular difficulties in manufacturing. Each of the four units was welded from twelve pie-shaped gores that were 13mm thick and 2.6m wide at the circumference, tapering down to less than 0.8mm where they met the central circular piece. To ensure that these were formed to the correct shape for the domes, North American

ABOVE Aft end of a displayed S-II to show the conical thrust structure. Note the ducts that route hydrogen around the LOX tank and along the structure to the engines. *(Mike Jetzer-heroicrelics.org)*

BELOW Diagram to show the location of the S-II heatshield. *(NASA)*

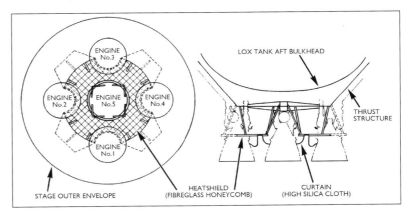

ABOVE Four cylinders and a forward bulkhead of a hydrogen tank are lowered onto two further cylinders already attached to the LOX tank and common bulkhead. Assembly at Seal Beach, California. *(NASA)*

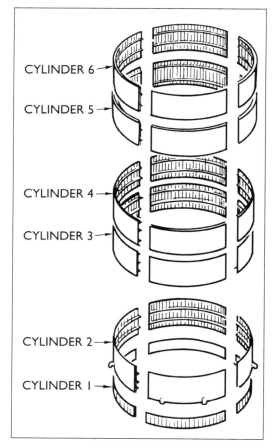

ABOVE Arrangement of wall panels to produce an S-II hydrogen tank. *(NASA/North American Aviation/Mike Jetzer)*

BELOW Drawing of a hydrogen tank cylinder panel to show the milled ribs on the inside. *(NASA/North American Aviation/Mike Jetzer/Woods)*

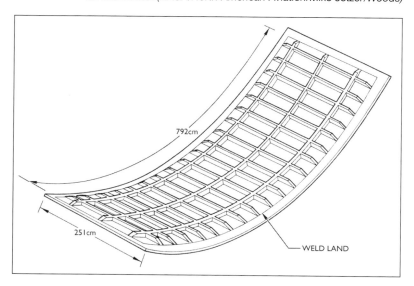

chose to use the pressure waves from explosive charges detonated underwater to form the individual pieces. A layer of insulation, 4cm thick, was added around the outside of the LH_2 tank. A common bulkhead separated the two tanks.

Because the LOX tank was domed outwards at both ends, the bottom of the LH_2 compartment was domed inwards. The five LH_2 outlets to the engines were therefore placed at the periphery, near to its lowest level. These fed into ducts which were routed outside the tank assembly's outer skin and around the LOX compartment, then along the thrust structure to the engines.

The LOX compartment, having a conventionally downwards facing dome at its base, had a single outlet at the very bottom that led into a sump. A cruciform baffle was mounted just above the outlet to minimise the

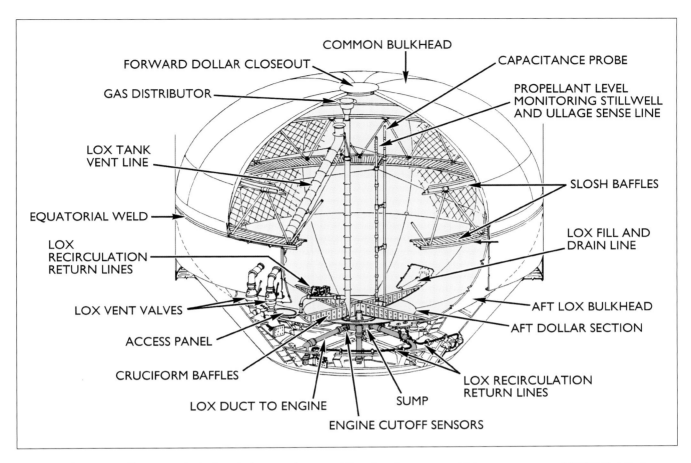

ABOVE Detailed cutaway drawing of an S-II LOX tank and LOX ducts. (NASA/Woods)

build-up of vortexes. Separate ducts then led from the sump to each of the engines. This arrangement minimised the amount of unusable LOX in the tank.

Common bulkhead

As a means of reducing the length, and therefore the weight of the S-II, North American opted to combine the two propellant tanks into a single unit with a common bulkhead between them. This design has the huge advantage of dispensing with the need for a heavy intertank cylinder and it was not an unusual feature of rocket systems. The venerable Agena upper stage, which flew throughout the 1960s, featured a common bulkhead as did the Centaur. As a counter example, the S-IC had an intertank to provide clearance between the domed bulkheads of its two tanks. On the other hand, the 200°C temperature difference between LOX and RP-1 made the advantages of a common bulkhead less desirable.

By significantly shortening the length of the stage, and therefore the space vehicle as a whole, the common bulkhead had beneficial cost implications for the entire launch support system all the way back to the design of the VAB.

The common bulkheads in the S-II and S-IVB stages had to deal with the temperature difference between the two cryogenic propellants of about 70°C. Without insulation, warmth from the LOX would increase the boil-off from the hydrogen fuel and the extreme cold of the LH_2 would freeze the oxygen, which is an unstable form of the element. In order to prevent the temperature of one liquid adversely affecting the other, the bulkhead needed to incorporate appropriate insulation.

However, a greater problem faced by North American was that the S-II's common bulkhead would be so huge that NASA had serious doubts whether the contractor could pull it off, given the mass of propellant on either side. The agency initially insisted that the company keep a design that involved separate bulkheads in reserve, lest it proved impossible to make a very large common bulkhead work.

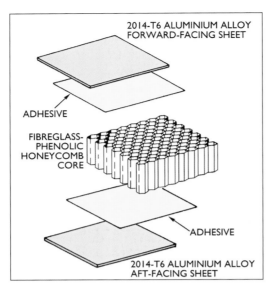

RIGHT Arrangement of layers within the common bulkhead. *(NASA/Woods)*

FAR RIGHT The surface of an early S-II hydrogen tank showing the cellular arrangement of the insulation, each cell filled with foam. *(Mike Jetzer-heroicrelics.org)*

North American eventually came up with a novel solution for the production of the bulkhead. It was formed from two separately made domed sheets of 2014 alloy. The lower sheet, which formed the upper bulkhead of the LOX compartment, was gently inflated to hold it in shape as a honeycomb insulator was applied on top and affixed with adhesive. This was then cured in a large autoclave.

A vacuum chamber was installed over the upper sheet, which was essentially the lower bulkhead of the LH_2 compartment so that the ambient air pressure would inflate the dome to the requisite shape when the chamber was partially evacuated. The whole arrangement was then brought onto the lower section and a profile taken to enable the insulation to be milled to achieve a perfect fit. By the time it was complete, this layer of insulation was 130mm thick at the centre and only 2mm at the edges.

Insulation

Being so light and so cold, liquid hydrogen is extremely volatile and will vaporise with very little encouragement. As well as insulation across the common bulkhead, the S-II's fuel compartment required extensive wall insulation to cope with a temperature difference of nearly 300°C between the daytime heat of Florida and the cryogenic contents in order to minimise the amount of fuel that would boil off before launch and during flight. Because the initial method chosen by the company proved troublesome, engineers had to revisit the problem part way through the programme.

At first, the fuel compartment was insulated with preformed panels that were glued to the outside of its cylindrical wall. This was in contrast to the S-IVB, where McDonnell insulated the inside of the fuel compartment. The advantage of the latter arrangement was that the insulation's adhesive wasn't being called upon to work at the temperature of liquid hydrogen since it was itself protected by the insulation.

However, North American's engineers were keen to use the extra strength gained from 2014 alloy when its temperature was brought down to 20K. This cryogenic strengthening would help to keep the stage as light as possible, which was a key goal, but in the process of devising a solution they overengineered it. Panels were created, 4cm thick, with a honeycomb core that was formed from phenolic resin. Each of the cells was filled with foam. Phenolic sheets were then glued onto either side of the honeycomb. Affixing these panels to the tank wall proved to be difficult because there would always be air pockets within the adhesive.

When a tank was chilled by fuelling, any air contained in the pockets between the wall and the insulation would liquefy and cause the panels to come away. To counter this, North American arranged for grooves to be cut in the panels next to the tank. Helium gas, which won't liquefy at LH_2 temperatures, was pumped through these grooves prior to and during the fuelling process in order to remove traces of air.

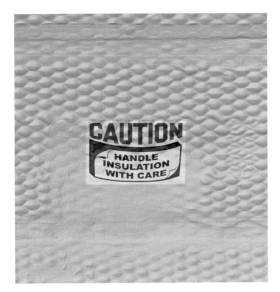

FAR LEFT Close-up of a warning sticker affixed to an insulation panel on an early S-II. *(Mike Jetzer-heroicrelics.org)*

LEFT The smoother surface of a later S-II insulated with spray-on foam. *(Mike Jetzer-heroicrelics.org)*

This cumbersome and expensive workaround had limited success.

From Apollo 13's launch vehicle (AS-508) onward, the S-II sported a completely different type of insulation; spray-on foam. This avoided the problem of air pockets because the foam's application directly onto the wall excluded them. Technicians merely had to spray more foam than required and trim off the excess to achieve an insulating layer that was cheaper, lighter and remained in place. It was a solution that was carried over to the external tank of the Space Shuttle.

A range of ancillary equipment was required to operate the stage properly. This included pipes and ducts to fill and drain the tanks, to purge them and to circulate both propellants through the engines to chill them and thereby ensure that only liquid, not gas, would pass through the pumps upon start-up. There were probes to measure the quantity of propellant within the tanks, valves to control their pressurisation, and instrumentation to sense the state of the machine and telemeter the resultant data to engineers on the ground.

Level sensing

The loading of propellant into the S-II and the timing of engine cut-off were achieved by reference to a suite of sensors. There were two types. Continuous capacitance probes gave a constant reading of the quantity of propellant within the tanks. Point level sensors essentially indicated whether they were wet or dry; i.e. immersed in liquid or surrounded by gas.

The capacitance probe was a tube within a tube arrangement that ran the length of the tank. Each tube was conductive and by being positioned adjacent to each other, they exhibited the electrical property called capacitance. This is their ability to store an electric charge. The capacitance of the probe was determined by whether the space between

BELOW Apollo 15's S-II awaits stacking in the VAB. The LH_2 ducts and the LH_2 fill and drain receptacle do not yet have their fairings attached. *(NASA)*

the nested tubes was occupied by the liquid propellant or the gas within the ullage space above. Capacitance therefore gave a measure of the level of the liquid in the compartment. From this it was straightforward to calculate the quantity. This measurement was used during the filling of the tanks and while the stage was operating in powered flight.

Each capacitance probe had two point level sensors near the top. The lower sensor indicated when the tank was nearly full. It acted as a backup means of indicating that the fast rate of loading the tank should be terminated and that the slower top-off rate should be implemented. The upper sensor indicated that the tank was about to be overfilled. If the liquid level reached this sensor then loading was stopped.

When the stage was part of an Apollo lunar mission, the timing of engine cut-off was based on the outputs from point level sensors. Each tank compartment had five sensors; ten in all. Those in the LH_2 tank were situated at each of the five outlets distributed around the periphery of the common bulkhead. The sensors in the LOX tank were clustered around the outlet to the sump located at the very bottom of the compartment. If any two sensors within a single tank were to indicate dry, this would be interpreted by the instrument unit controlling the flight that the stage was almost empty and that it was time to shut down the engines. By leaving the least possible unburned propellant in the tanks, the stage could maximise the impulse delivered to its payload.

To ensure that false indications from these cut-off sensors didn't prematurely terminate the S-II's powered flight, the level sensing system that translated their output into a command wasn't armed until the reading from the capacitance probe indicated imminent propellant exhaustion. This way, should a false indication cause S-II cut-off to occur a few seconds early, it would be late enough to enable the margin in the S-IVB to make up the shortfall and still continue the mission.

Propellant utilisation

In the wake of demands by NASA for the S-II to perform better, one strategy to squeeze the most out of the stage involved a means to maximise propellant utilisation.

In its role of lifting an Apollo lunar spacecraft the S-II wasn't required to shut down at a particular velocity. Rather, it was called upon to burn until its propellant was very nearly exhausted, having achieved about 90% of the speed required for orbit. Any propellant that remained in the stage at shutdown had cost dearly because, not only had it been accelerated to fantastic speeds, it also did not then contribute to the stage's further acceleration. Careful design of the tank outlets helped to minimise the amount of unused propellant, but the key thing was to ensure that the fuel and LOX both became depleted in such a way as to maximise efficiency.

At first glance, it would appear that all that was required would be to ensure simultaneous depletion of the propellants. But the inherent variables would mean that there was an equal chance of some LOX or fuel remaining. Since LOX is much heavier than hydrogen, it was therefore preferable for LH_2 to be left remaining. Ground technicians therefore attempted to bias their fuel loading with an additional 250kg or so of LH_2.

But there was more. The exact rate at which each engine consumed propellant differed slightly between units. This depended on the mixture ratio being burned and the flowrates of propellant through their ducts. Moreover, owing to the continuous boil-off and topping off of the cryogenic liquids there was always some uncertainty about the exact quantities of propellant available for combustion. S-II propellant continued to boil off during the S-IC's flight, further muddying the waters when it came to understanding how much would be left at engine cut-off.

The trick that the engineers used was the fact that the J-2 engine could have its mixture ratio (MR) changed during operation. Indeed, the valve that achieved this was often known as the *propellant utilisation* valve. What it did was to partially bypass the engine's oxidiser pump, thereby reducing its ability to pump LOX into the combustion chamber.

Early in the programme, the strategy used a closed-loop control system. By using the measurements of propellant quantity from the tank capacitance probes, the computer in the instrument unit could monitor the rate of depletion of the propellants in real time and decide when to change from a high to a low MR. It worked, but there were bigger gains to be made.

Mathematical analysts noted that, given the complexities of all the variables that affected residual propellant, they were more likely to achieve simultaneous depletion if they simply switched to the lower MR when the vehicle had reached a pre-determined velocity. This was an open-loop method because the system wasn't feeding back information from itself in order to make its decision. This scheme was used from Apollo 8 (AS-503) onwards.

Tank pressurisation

Two techniques were used to pressurise the S-II's propellant tanks. Through to the last few minutes before lift-off, both tanks were being continuously topped off with propellant as the leakage of heat from outside caused the cryogenic liquids to constantly boil. The resultant vapour was allowed to vent from the tanks, with the oxygen being released directly into the atmosphere, its cold temperature forming distinctive white plumes that trailed away from the vehicle in the wind. The hydrogen vapour had to be handled carefully owing to its flammability. It was routed 300m away from the launch pad to a large pond of water, 31m × 28m and 80cm deep, where it bubbled up to the surface and was safely ignited by an arrangement of hot wires.

Three minutes prior to launch for the LH_2 tank, and 1½ minutes for the LOX tank, the valves that controlled the venting, two per tank, were closed. Further boiling of the propellants helped to raise the pressures in the tanks, but full pressurisation was achieved by helium gas fed from a supply on the launch pad. This ensured that both tanks were fully pressurised by the time the 30-second point was reached in the countdown.

The vent valves were configured as relief valves so that neither tank on the S-II would exceed the allowed pressures. The LOX tank was limited to 290kPa (42psi), but the LH_2 tank was limited to 203kPa (29.5psi) during S-IC operation and 228kPa (33psi) during S-II powered flight.

LEFT A J-2 propellant utilisation valve. On the J-2 these were used to adjust the consumption of LOX and thereby ensure near simultaneous depletion of the S-II's propellants.
(Courtesy of Scott Schneeweis collection – spaceaholic.com)

After the S-IC had done its job and the S-II engines had taken over, pressurisation was carried out in conjunction with the J-2 engines, using their heat to generate the necessary gases. To pressurise the LOX tank, a small amount of the oxidiser was passed through the heat exchangers on the J-2s. The hot gas that had powered the turbopumps would warm the LOX and turn it into gas. This was then piped to the ullage space at the top of the LOX tank to maintain its pressure.

In the early S-IIs, this feed included a pressure regulator to control the pressure of the gas being fed to the tank. As their insight into how the stage operated increased, engineers replaced the complex regulator with a simple calibrated orifice to set the pressure, starting with Apollo 15 (AS-510). It saved additional mass in the stage, thereby raising the vehicle's payload, and it improved overall reliability. Should the tank pressure rise too high, the vent valves would open to restrict it to the design pressure.

The fuel tank was also pressurised by its own contents, but in this case there was no need to have a heat exchanger generate gaseous hydrogen (GH_2). The J-2 engines were already producing copious quantities within the pipework of their thrust chambers because the liquid fuel was being vaporised prior to being injected into the combustion chamber. It therefore required a small amount of this gas to be bled off from the injector manifold of each engine. This was then fed to another manifold that gathered the gas from all five engines and sent it along a duct to the top of the LH_2 tank, where it entered via a distributor.

ABOVE Apollo 12's S-II being lowered onto the S-IC stage in the VAB. *(NASA)*

As was done on the LOX pressurising system, later stages replaced the pressure regulator with a calibrated orifice within the feed to regulate its feed pressure, relying on the tank's vent valves to cope with any overpressure.

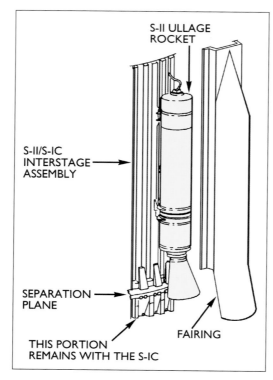

RIGHT Drawing of an S-II ullage rocket mounted on the interstage ring. *(NASA/Woods)*

Propellant dispersion

While the S-IC was pushing the vehicle through the atmosphere, safety considerations required a means of dispersing the S-II propellant in the event of an abort, especially if the vehicle had lost control. This was true for all three stages. Rather than allowing them to impact the ground fully loaded, it was felt better to break up the stages while the vehicle was still at altitude.

Linear charges placed along the length of the LH_2 tank and around the LOX tank would quickly open up both tanks and allow their contents to be dumped into the atmosphere with minimal mixing. Of course, this activity would only occur after the launch escape tower had pulled the spacecraft free.

Soon after the S-II began its powered flight and the escape tower was jettisoned, this pyrotechnic system was turned off, or *safed*, because the vehicle had reached a distance out over the ocean and was at an altitude where it could no longer impact on populated areas.

Ullage rockets

When the S-II was designed, the interstage ring between it and the S-IC included eight powerful solid-propellant rockets to push the S-II forward as soon as the S-IC had been jettisoned. At that moment, the unpowered vehicle was coasting for a short time through the upper atmosphere and therefore essentially weightless.

The ullage rockets were housed within aerodynamic fairings and they did two things. They forced any gas voids within the propellant tanks to move away from the tank outlets, thereby ensuring that the ullage space was at the top. They also created an additional head of pressure at the engine inlet over and above that provided by the pressure within the tanks.

The ullage rockets were one of those systems whose necessity was later reassessed. Only the first unmanned test flight of the S-II had eight ullage rockets. All subsequent missions up to Apollo 14 (AS-509) included four of the units. Interestingly, although Apollo 6 had fairings for eight rockets four of them were empty.

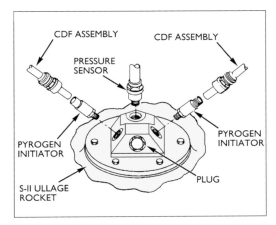

ABOVE Details of the ignition system for the S-II ullage rockets. *(NASA/Woods)*

By Apollo 15, engineers were pushing at the Saturn V's margins in order to increase the mass of payload that could be landed on the Moon. One way to improve the capability, starting with that mission (AS-510), was to delete the S-II ullage rockets.

Each ullage rocket consisted of a cylindrical steel case, 0.31 × 2.26m, filled with 150kg of solid propellant. A nozzle at the end directed the 100kN thrust 10° away from the stage's centreline so that if one unit were to fail, the offset thrust arising from the rocket on the opposite side would be less prone to making the vehicle rotate. This was because the thrust would be aimed nearer to the stack's centre of mass.

The command to ignite the ullage rockets came one-fifth of a second before the first stage was jettisoned. They fired for about 4 seconds and were discarded with the interstage ring about 30 seconds later.

A singular stage

In the final analysis, the S-II stage, despite its difficult gestation, proved to be a reliable workhorse right through the Saturn V programme. Although it tested the will of NASA and its contractors, it worked when it needed to. The experience gained from the fabrication of its huge cryogenic tanks proved invaluable when it came to building the external tanks that fuelled the Space Shuttle for a generation.

ABOVE An S-II stage is lifted into the A-2 test stand at the Mississippi Test Facility. *(NASA-MSFC)*

BELOW A test firing of an S-II stage at the Mississippi Test Facility. The exhaust product from the stage was superheated steam. *(NASA-MSFC)*

99

S-II: THE TROUBLED STAGE

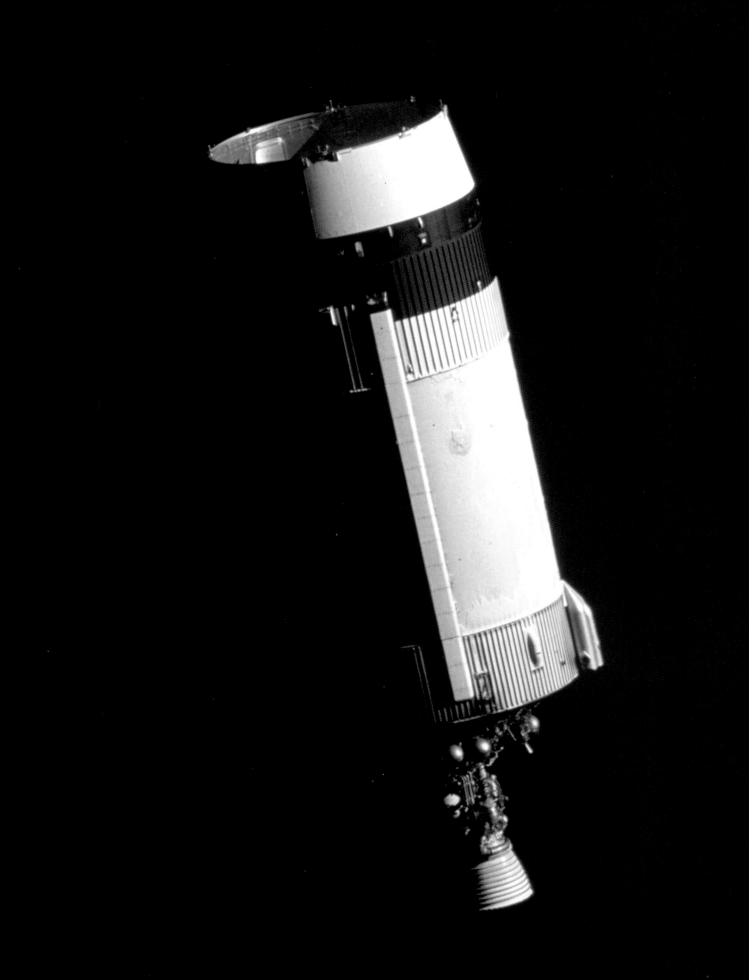

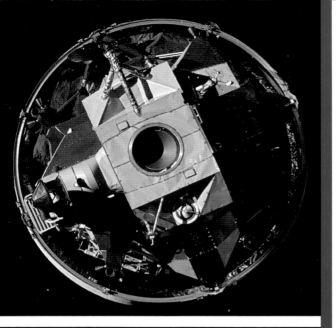

Chapter Six

S-IVB: A stage to the Moon

The S-IVB's place in the Saturn V story is especially significant, thanks to its dual role in an Apollo lunar mission. The first was to take its turn after the other two stages in helping heft a complete Apollo spacecraft into low Earth orbit, shutting off at a precise velocity and direction to ensure at least two safe revolutions around the planet. But it was its second task in a mission that really set it apart from any contemporary rockets, for it was the S-IVB that was reignited to set humans on a path to the Moon.

OPPOSITE Apollo 17's **S-IVB** coasts alongside the spacecraft as both hurtle to the Moon. Its fate was to impact the lunar surface for seismology. *(NASA)* **Top:** Apollo 15's lunar module, *Falcon*, still cradled in the top of its **S-IVB**. *(NASA)* **Middle:** The forward bulkhead of Apollo 9's **S-IVB** after LM extraction. *(NASA)* **Bottom:** Apollo 8's **S-IVB** with a toroidal water tank to simulate the mass of the **LM**. *(NASA)*

ABOVE A Saturn S-IV stage intended for the SA-9 vehicle (see image on page 21). Propulsion for the stage was provided by six RL-10 engines burning hydrogen and oxygen. *(NASA-MSFC)*

The S-IVB was the workhorse of the Apollo programme because of its additional role as the second stage of the Saturn IB vehicle until the mid-1970s. In an additional flourish and in a triumph of adaptation, a single example was launched into Earth orbit in 1973, having first been completely remodelled to become the Skylab space station and a temporary home for three crews of three astronauts.

The S-IVB was a derivative of an earlier Saturn stage, the S-IV, which was the first hydrogen-powered rocket to come out of the Marshall Space Flight Center. The S-IV was conceived when NASA was still trying to work out what shape the Saturn family of launch vehicles should take. The initial intention was for the S-IV to be a fourth stage for one of its paper designs, the C-4.

Originally, the S-IV was to be powered by four Pratt & Whitney LR-119 units which were then in development. But when the C-4 arrangement was dropped, the S-IV morphed into a second stage for the C-1 vehicle. A further change came about when delays in developing the LR-119 engines provoked NASA to have the stage powered by a cluster of six of the already established RL-10 motors that were being used in Convair's smaller Centaur upper stage.

The contract for the S-IV stage went to the Douglas Aircraft Company in 1960 (later McDonnell Douglas after a merger). Douglas was chosen over Convair even though the company was new to hydrogen propulsion technology. It was, however, a veteran rocket builder having produced the Thor missile. NASA wished to spread the country's experience in this exotic field, given that Convair were already busy with the Centaur.

As NASA's engineers pondered the tasks that their super-booster, the C-5, might be called upon to achieve, they came up with the concept of an uprated S-IV that would be the vehicle's third stage. Using a single example of the new human-rated J-2 engine, this stage would be powerful enough to send an Apollo spacecraft towards the Moon. At 5.6m diameter, it was the first definition of what became known as the S-IVB.

It was realised, however, that as well as acting as a booster to take an Apollo spacecraft out of Earth orbit, the S-IVB could fulfil other tasks. Managers had perceived a need to test Apollo components in Earth orbit. Rather than use an entire C-5 just to get the Apollo spacecraft aloft for testing, they created an uprated version of the C-1 that could do the job

RIGHT S-IVB-206, a version of the stage used on the Saturn IB vehicle. This S-IVB, seen here at KSC, became the second stage of the first manned Skylab mission, launched on 25 May 1973. *(NASA-MSFC)*

with just two smaller stages. This new vehicle, initially called the C-1B, was later renamed the Saturn IB and it required that the S-IVB be made even more capable. The change caused its diameter to be increased to 6.6m, which was very slightly larger than the S-IB first stage.

By using this larger, more capable S-IVB on the C-5, NASA also raised the payload that could be sent to the Moon. This led to two variants of the S-IVB, largely identifiable by the interstage that acted as an adapter to the rest of their vehicle. The unit for the C-1B was a cylinder while that for the C-5 was a large cone to join the S-IVB to the 10m diameter S-II stage.

Thus the S-IVB gained dual roles. For Earth orbit missions, it could insert an Apollo command and service module into orbit and then, if necessary, become a target to enable astronauts to practise the manoeuvres required to pluck a lunar module that could be mounted at its top. Alternatively it could place an unmanned lunar module into orbit to conduct tests of that spacecraft. For lunar missions, it would finish off the job of lifting an entire Apollo stack, including a lunar module, into orbit. Then, for a few hours spanning several orbits, it would keep itself stabilised while systems were checked out. It would then relight and send the spacecraft on a lunar trajectory. The numbering system for the stages indicated their role: the 200-series were intended for the Saturn IB while the 500-series indicated that an S-IVB was for the Saturn V.

NASA intended for the new stage to be heavily based on work already completed by Douglas for the S-IV, itself having taken advantage of the company's experience building the Thor missile. This led to Douglas

RIGHT S-IVB-505N being removed from Beta Test Stand 1 at Sacramento after its acceptance tests. This stage became part of Apollo 10 and is still in orbit around the Sun. *(NASA-MSFC)*

ABOVE An S-IVB being unloaded from an early version of the Super Guppy aircraft at Redstone Airfield, Huntsville. This stage, S-IVB-212, would be converted to become the orbital workshop core of Skylab. *(NASA-MSFC)*

RIGHT A series of still frames from 16mm footage showing the destruction of S-IVB-503 at Sacramento on 20 January 1967. *(NASA/Woods)*

being awarded the S-IVB contract without competition. Final assembly of the stage would be undertaken at their existing plant at Huntington Beach, California, and completed stages would be shipped by air using outsize propeller-driven Super Guppy aircraft originally converted from C-97J Turbo Stratofreighters.

Though Douglas had an easier time with the S-IVB than North American had with the S-II, the gestation of the S-IVB was not entirely without incident. The most notable mishap occurred in January 1967, the darkest month of the entire Apollo programme. On 20 January, only one week before three Apollo astronauts would die in the Apollo 204 fire at the Kennedy Space Center, S-IVB-503 was being prepared for a test firing at Douglas's Sacramento test facility in California.

NASA's numbering system was consistent across all three stages for a particular Saturn V, so that the last digits would be the same as for the final space vehicle in which they were to be flown. For example, Apollo 15 was launched as AS-510, a complete space vehicle. Its rocket was SA-510 and comprised S-IC-10, S-II-10, S-IVB-510 and IU-510. It followed that S-IVB-503, about to be tested on the Beta III test stand, would be part of the AS-503 Saturn V which eventually sent the Apollo 8 spacecraft to the Moon. But it never got that far.

As S-IVB-503 sat in the stand, fully fuelled and pressurised, a countdown was being rehearsed that would lead to a simulated lift-off by a non-existent first stage, followed by an equally virtual second stage. As far as -503 was concerned,

it was being fed the same fluids and signals it would receive on a real flight. When its turn came, about 8½ minutes in, it was to ignite for real and exercise the systems that it would use in flight.

Eleven seconds prior to the simulated lift-off and 511 seconds before it was due to ignite, S-IVB-503 suddenly and violently exploded, destroying itself as the two volatile propellants were quickly introduced to each other.

Engineers scoured the area for debris and found half a helium tank in the flame trench, not too far from the remains of the S-IVB. This was a hemisphere from one of the eight spherical tanks that were mounted adjacent to the engine. These stored helium at the very high pressure of 22MPa (3,200psi).

What distinguished this hemisphere from the others that were found was that, unlike them, it had not been blackened by the heat of the conflagration, signifying that it had come apart prior to the explosion. Its other half was never found and studies showed that the wrong welding filler had been used to join the hemispheres together. Although titanium alloy had been specified, a pure titanium filler had been used. The weld had managed to hold as -503 passed through repeated tests until this occasion, when the release of pressure caused one hemisphere to rip through the propellant tanks, igniting them. It was a shocking reminder of the care needed when building machines that had to work at the very edge of their capabilities.

Despite the loss of -503, NASA kept to their naming scheme. Thus -504 became S-IVB-503N, the 'N' indicating that it was a 'new' -503. This numbering jump continued until -506 became -505N, that being the last of the stages in the system at that time.

Airframe

At first glance, the structure of the S-IVB appeared to be very similar to the S-II. A common insulated bulkhead separated the storage volumes for liquid oxygen (LOX) and liquid hydrogen (LH_2) within what was essentially a single large tank assembly. Each end of this assembly was attached to a cylindrical skirt to accommodate the tank's domes. The engine was mounted on a conical thrust structure fastened to the aft end of the tank assembly. It was then housed within an interstage

ABOVE The physical aftermath of the destruction of S-IVB-503. *(Courtesy of Jim Porter/Terri Pennello and Alan Lawrie)*

BELOW One half of the helium storage tank that came apart due to a faulty weld. The other half shot through S-IVB-503, igniting its propellants and destroying it. *(Courtesy of Jim Porter/Terri Pennello and Alan Lawrie)*

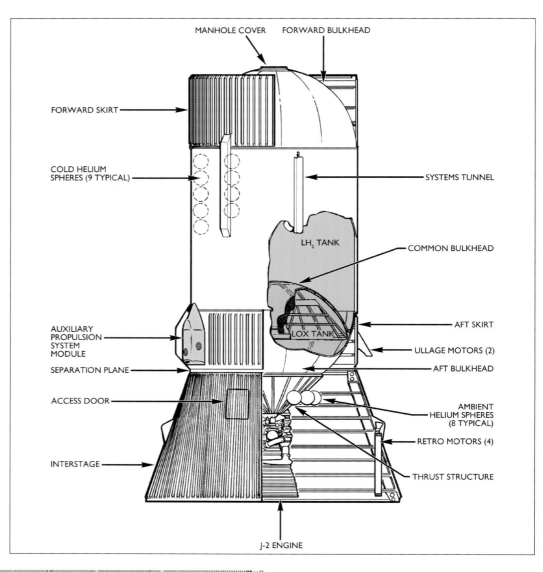

RIGHT Cutaway drawing of the Saturn V version of the S-IVB, indicating its major components. *(NASA/Douglas Aircraft Company/Mike Jetzer/ Woods)*

BELOW Front to rear, the forward skirt, aft skirt and thrust structure for S-IVB-507, the Apollo 12 third stage. Behind them is S-IVB-211. *(Courtesy of Phil Broad)*

component that sat between the aft skirt and the S-II below.

Each skirt was a cylinder fabricated from 7075-T6 aluminium alloy and stiffened with hat-section longitudinal stringers on the outside and ring frames on the inside. These provided the structural support necessary to mate the tank assembly to the rest of the Saturn V; the 3.1m-long forward skirt being attached to the instrument unit above and the 2.1m-long aft skirt being attached to the conical interstage below. Each included a mount for the various items of ancillary equipment that enabled the stage to do its job; i.e. electronics, batteries, antennae and environmental control units. The aft skirt also carried two ullage rockets and two auxiliary propulsion packages.

RIGHT A technician works on the thrust structure for S-IVB-509, the Apollo 14 third stage. *(Courtesy of Phil Broad)*

CENTRE Cutaway drawing of the S-II/S-IVB interstage. *(NASA/Douglas Aircraft Company/Mike Jetzer/Woods)*

The thrust structure was also made from 7075 alloy and formed a cone that carried the stage's single J-2 engine mounted at its narrow end. Stiffened with longitudinal hat-section stringers and rings, it served to evenly spread the force from the engine to the bulkhead and onto the tank's cylindrical wall. It provided support for a fuel line to the engine and at least eight titanium spheres filled with helium as well as other services to the engine.

Although it was manufactured with the S-IVB, the interstage section was discarded along with the S-II stage. It adapted the 6.6m diameter of the third stage to the 10m diameter of the second stage and it provided clearance to house the single engine. An access door allowed technicians to enter the space during preparation and it housed four solid-fuelled retrorockets that helped separate the stages. It was also produced using 7075 alloy with stiffening stringers.

Tankage

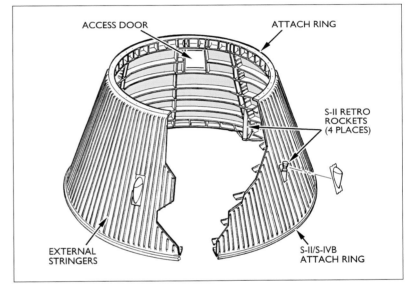

The tank assembly was 13.4m long and 6.6m in diameter and had two compartments to contain the propellants. At the bottom was a smaller volume that held 77,000 litres of LOX. Above it, the LH_2 section held 286,000 litres.

Its form was very similar to the S-II with two end bulkheads, a cylindrical middle section, and an internal common bulkhead dividing the two tank compartments. Whereas the S-II used ellipsoidal bulkheads for the tank assembly in order to keep the stage's length and hence mass down, the ends of the S-IVB were true

RIGHT The wall of the hydrogen tank for S-IVB-514. This was welded from eight panels, each with a pattern of ribs milled into them for strength. 514 now hangs at the Apollo Saturn V Center in Florida. *(Courtesy of Phil Broad)*

ABOVE The common bulkhead for S-IVB-506, the Apollo 11 third stage. *(Courtesy of Phil Broad)*

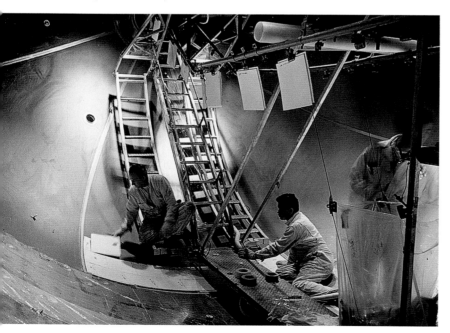

BELOW Insulating tiles being laid into the forward bulkhead of S-IVB-506, the Apollo 11 third stage. *(Courtesy of Phil Broad)*

hemispheres of 3.3m radius. Douglas accepted the length penalty in return for the extra strength that this configuration provided.

The structure used the same 2014 aluminium alloy that North American would use to build their S-II tank assembly. Douglas's method of stiffening the external walls of the LH_2 tank differed from the rectangular internal strengthening used for the S-II. For the S-IVB, rectangular recesses were milled into sheets of 1.9mm thickness to leave behind a waffle-like pattern of strengthening ribs orientated at 45° and with a pitch of 24.13cm (precisely 9.5 inches). These sheets were then formed to the proper curve before being welded together.

The common bulkhead that separated the tank compartments had the same spherical form with a 3.3m radius as the end bulkheads but its diameter was lessened by virtue of being set slightly into the aft hemisphere in order to accommodate the required quantity of LOX. There was no need for the lower compartment to be a full sphere.

Two domed sheets formed the common bulkhead. The lower sheet that faced into the LOX tank was 1.4mm thick and the upper sheet, facing into the LH_2 tank, was only 0.813mm thick. Between them was 5cm of insulation formed from a fibreglass honeycomb stiffened with phenolic resin.

Insulation

Unlike the external insulation employed by North American on the S-II, Douglas didn't rely on the strengthening properties of 2014 alloy at cryogenic temperatures. Instead, they chose to install their insulation on the inside of the LH_2 tank's external walls so that the metal skin wouldn't feel the full chill of the contents' 20K temperature. This scheme also avoided exposing the adhesive to very low temperatures; a factor which bedevilled North American's scheme.

The S-IVB insulation was constructed from thousands of polystyrene foam blocks, each specifically shaped to fit the waffle pattern on the inside of the tank walls. Once they had been glued into place, their interior surface was sealed by being lined with fibreglass cloth that had been impregnated with resin.

With the insulation installed, other fittings required within the tank were added, such as level sensing devices, helium storage spheres, and the valves and pipework to fill and empty the tanks and control their internal pressure.

Level sensing and propellant utilisation

Each tank contained a full-length tube-within-a-tube capacitance probe to measure the quantity of propellant remaining. These sensors aided the control of the loading process and allowed levels to be monitored during the stage's operation. Additionally, a set of point level sensors at the top and bottom of the tank augmented this system and indicated when the tank was nearly full and when it was approaching depletion.

Initially, the intention had been to employ the output from the capacitance probes to help to ensure maximum depletion of the propellants. This would have used the ability of the S-IVB's single J-2 engine to adjust its mixture ratio during combustion. As the levels fell, the Saturn's computer would calculate the best time to switch from burning a fuel-rich mixture (4.5:1, LOX to LH_2) to the normal mixture of 5:1 and thereby alter the relative rate at which the two propellants were being consumed. This closed-loop strategy was replaced early in the programme by an open-loop method that better reflected the job the S-IVB was being called upon to achieve.

Whereas the S-II's task was to burn almost all its propellant to exhaustion to ensure the stage had delivered as much impulse as possible, the S-IVB's task was to burn until it had attained a precise velocity, either to insert itself into the intended orbital path around Earth or, on a subsequent burn, to send the spacecraft towards the Moon on a trajectory that had to be nearly perfect. This meant that when the engine was shut down after the second burn there would be a significant quantity of propellant in the tanks.

Rather than worrying about whether every last drop of propellant would be consumed, the factor to be considered was the prodigious rate at which the LH_2 boiled away. What therefore concerned the engineers was whether there would be sufficient hydrogen fuel on board if the stage had to remain in Earth orbit not just for the planned time of about 2¾ hours, but also for an additional 90-minute revolution before being reignited for the translunar injection (TLI) burn.

The profile for an Apollo lunar mission required that the S-IVB, with Apollo spacecraft attached, coast in Earth orbit for not quite two revolutions. Throughout this time, as the crew checked their spacecraft, heat from the Sun beat onto the skin of the stage and leaked into the tank. This warmth caused the extremely volatile cryogenic fuel to vaporise and boil off and, as a result, a substantial amount of LH_2 was lost. This had to be taken into account when filling the tank.

However, NASA wanted to be able to make an extra contingency orbit in case the crew or mission control identified a problem during their checks that would need a little extra time to troubleshoot. During these extra 90 minutes,

ABOVE S-IVB-506 being hoisted through the VAB to be mated to the S-II of the Apollo 11 launch vehicle. *(NASA)*

even more LH_2 would be lost through boil-off and the delayed TLI burn would have less fuel available in comparison to an on-time burn.

The solution was to load enough fuel so that, even if it had to coast for the extra orbit, the S-IVB could complete both its first and second burns at a mixture ratio of 5:1 (LOX to LH_2). Hence for a normal mission there would be an excess of LH_2. This would be consumed during the TLI burn by running with a fuel-rich mixture ratio of 4.5:1 for the first few minutes, then using the propellant utilisation valve to switch to a ratio of 5:1 for the remainder of the burn. By design the J-2 engine provided a great deal of operational flexibility.

Engine prechill

Prior to the command to start the J-2 engine, it was necessary to ensure that the fuel and LOX lines leading to the turbopumps, as well as the pumps themselves, had been

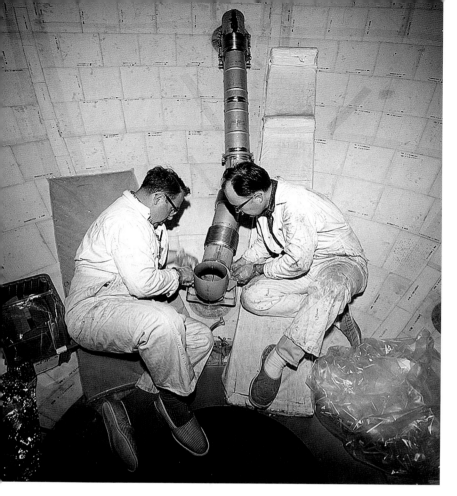

ABOVE **Installation of the vent and fill duct within the insulated hydrogen tank of S-IVB-505N.** *(Courtesy of Phil Broad)*

Tank pressurisation and venting

The operation of the engine's turbopumps depended on propellants being fed to their inlets at the correct pressure, this being set by the pressures in the respective tank compartments. Maintaining pressure became a balancing act between various factors acting to change it: the emptying of the tanks would tend to reduce it; the boiling of the propellants, especially the LH_2, as heat leaked in would tend to increase it. It was the responsibility of various systems to counter these influences at the appropriate times.

When propellants were being loaded and boil-off was prodigious, valves were kept open to vent oxygen to the air and to route hydrogen to the burn pond some distance away for safe combustion.

Shortly before launch, the vent valves were closed and helium was pumped in from a ground-based source to pressurise the LOX tank to 283kPa (41psia) and the LH_2 tank to 214kPa (31psia). While the lower stages were doing their work and the propellants in the S-IVB continued to boil off, relief valves operated as necessary to prevent overpressurisation. In some cases, extra helium was added to maintain pressure.

When the J-2 began its first burn the pressurisation regime changed to compensate for the declining fluid level in the tanks. A major difference between the S-II and S-IVB arrangements was that in the former, the J-2's heat exchanger warmed LOX to create gaseous oxygen that pressurised the LOX tank. In the S-IVB, LOX pressure was maintained between 262 and 283kPa (38 and 41psia) by heating helium in the engine's heat exchanger and feeding that to the tank. Since this helium was supplied from a set of storage spheres within the LH_2 tank, it was extremely cold prior to being passed through the heat exchanger.

AS-501 and -502 (the Apollo 4 and 6 boosters) had eight helium spheres in the tank. The remaining vehicles had nine. Additional helium was available from storage spheres mounted on the exterior of the thrust structure but being at ambient temperature, they couldn't hold as much helium as the cold spheres.

The method of pressurising the LH_2 tank was the same as for the S-II, in that gaseous

brought down to the same temperature as the propellants so that gas would not enter the pump during start-up. Failure to do so would allow the pump to race, possibly damaging it.

While fuel and oxidiser were being loaded into the tanks, the prevalves in the propellant ducts that led to the engine were kept open to allow the cold liquids to enter the turbopumps and begin the process of lowering their temperature.

Then, five minutes prior to lift-off, the prevalves were closed and two recirculation systems were activated. These maintained a constant flow of propellant from each tank via separate lines to the top of the respective duct just below its prevalve. The circulation continued down the duct, through the respective turbopump to return via other lines to the tanks. Valves controlled this circulation and it was maintained by electrically operated pumps within each system.

This preconditioning circulation was maintained through launch and the flight of the lower stages, ceasing just prior to the start of the S-IVB's J-2 engine. Towards the end of the orbital coast, it was resumed for about five minutes, ceasing shortly prior to the TLI burn.

hydrogen was tapped from the engine as it entered the injector manifold and was fed to the top of the fuel tank via a control module. A valve controlled the flow of this gaseous hydrogen based on the pressure in the tank, maintaining it between 193 and 214kPa (28 and 31psia).

Between entering Earth orbit and making the second burn, the stage had to contend with the Sun's heat which continued to leak into the LH_2 tank. A pair of valves came into operation to overcome this ongoing boil-off by allowing the gas to vent and maintain the tank at between 134.5 and 145kPa (19.5 and 21psia).

This excess gas was routed to two vents that were deliberately aimed aft so that they provided a very small but constant thrust to the stage. Because a set of auxiliary thrusters (see page 113) were keeping the stack aimed in a sharp-end-forward attitude, the effect of these so-called propulsive vents was to maintain a gentle force on the propellants to keep them settled at the bottom of their tanks until preparations began for the second burn.

About 9 minutes before the TLI burn, commands from the instrument unit began the process of repressurising the tank compartments in readiness to deliver propellant to the J-2. Central to this was a small unit, the O_2/H_2 burner, attached to the thrust structure that gave warmth to extremely cold helium gas.

This O_2/H_2 burner was fed with propellants which were ignited by a spark to produce heat within a combustion chamber. As the exhaust exited via a nozzle, some of its heat was given up to a set of coils. These coils were to vaporise the incoming propellant prior to burning and also to heat helium supplied by cold spheres within the LH_2 tank. The warmed helium was then fed to the tanks to raise their pressure back to operational levels.

The nozzle of the O_2/H_2 burner generated a small but not insignificant thrust of about 170N (39lb-f). To account for this, and to avoid the force causing the stack to rotate, the burner's thrust axis was aligned with where the stack's centre of mass was expected to be at this part of the mission. By this arrangement, any rotation would be so small that the S-IVB's attitude control thrusters would have little difficulty countering it. Also, although the thrust would translate the stack by a small amount, at this stage of the mission it was of no consequence.

For the last minute prior to the TLI burn, both tanks were switched to being pressurised with helium fed from the ambient tanks as a contingency in case of a failure of the burner. Throughout the TLI burn, the stage resumed pressurising the tanks using the J-2's heat exchanger as it had done before.

Once the TLI burn had come to an end, a lunar-bound S-IVB had two major tasks remaining. The first was to hold itself at a predetermined attitude while the lunar module was extracted from the conical adapter section above the instrument unit. The criterion for this attitude was to provide a suitable lighting angle to aid the command module pilot as he carried out a docking manoeuvre.

With the spacecraft gone, engineers in mission control sent commands to the S-IVB to depressurise the tanks by completely venting or dumping their contents, particularly the LOX. However, in order to achieve one final mission goal, they used this dump as a source of propulsion to steer the stage either into solar orbit or to crash onto the lunar surface to provide a calibrated seismic signal for seismometers already in place. LOX dumping was achieved by opening the engine's LOX valves and allowing the propellant to exit via the

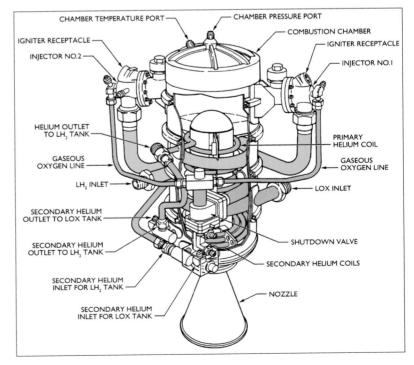

BELOW Drawing of the O_2/H_2 burner to indicate the routes for fuel, LOX and helium. *(NASA/Mike Jetzer/ Woods)*

ABOVE Installation of an O_2/H_2 burner onto the thrust structure of S-IVB-505. *(Courtesy of Phil Broad)*

BELOW The J-2 engine could be swivelled up to 7° from the stage's centreline. This diagram shows how deflection of the thrust vector away from the centre of mass could provide pitch control during powered flight. *(NASA/Mike Jetzer/Woods)*

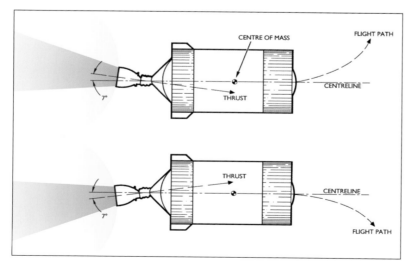

engine nozzle. The helium spheres were also vented at this time.

Attitude control

By virtue of possessing just a single main engine, the S-IVB design had a drawback when compared to the other two stages of the Saturn V. All three stages could perform pitch and yaw manoeuvres in powered flight by swivelling their gimballed engines. This *thrust vector control* form of steering is achieved by altering the aim of the engines. The S-IC and S-II could perform a roll manoeuvre by deflecting their outer engines to generate a corkscrew effect, but the S-IVB, having just one engine, couldn't do this.

The S-IVB's thrust vector control was carried out by the operation of two hydraulically operated servoactuators, each of which could deflect the engine up to 7° from the stage's centreline. Their hydraulic power was provided by either a primary pump driven off the J-2's LOX turbopump or a secondary electrically driven pump.

Because the servoactuators were closed-loop, when they were commanded to a particular position they would give mechanical feedback to the servovalve to indicate they had reached the required position and thus

BELOW Drawing of the location of the S-IVB's pitch actuator. This, and another for yaw, gave the stage 2-axis attitude control during powered flight. *(NASA/Douglas Aircraft Company/Mike Jetzer/Woods)*

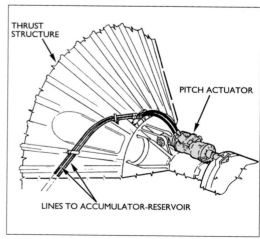

RIGHT Diagram of an APS module from the S-IVB. The major components are indicated. *(NASA/Woods)*

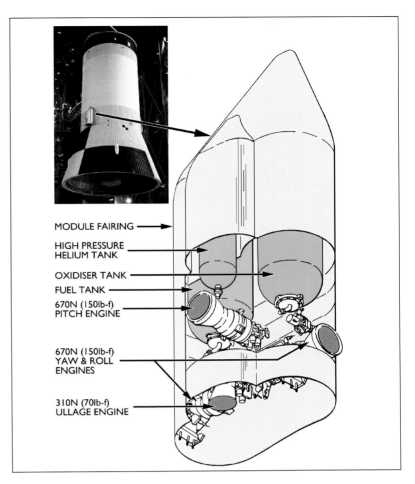

terminate the hydraulic pressure that was actuating them. Consequentially, their deflection was always proportional to the control signal coming from the instrument unit.

Although this arrangement controlled the pitch and yaw of the vehicle, it could only do so while the engine was operating. But the S-IVB's mission required it to maintain full 3-axis attitude control both during the 2¾ hours that it was coasting in Earth orbit and after the TLI burn when it had to hold a specific attitude while the crew retrieved the lunar module.

Auxiliary propulsion system

To provide roll control during powered flight and 3-axis control during coasting flight, the S-IVB included a pair of modules called the *auxiliary propulsion system* (APS) installed on either side of the aft skirt. Each of these modules contained four small rocket motors along with tanks for liquid propellants and pressurisation gas. Being self-contained, they could easily be replaced if preflight testing identified a fault.

A small motor of 310N (70lb-f) thrust was aimed to the rear of the stage. Its only function was to perform an ullage burn as soon as the stack entered Earth orbit, and again immediately prior to the second ignition of the J-2 engine for the TLI burn. The thrust from these two small engines would settle the S-IVB's propellants to the aft of the tank compartments.

The three other motors each provided 670N (150lb-f) of thrust. Two were aimed to either side to control yaw and roll rotation. The third was aimed perpendicular to the stage's profile to control pitch. The six large motors distributed between these two modules enabled the instrument unit to provide full 3-axis attitude control.

The APS was the only propulsion system on

RIGHT Given the toxic nature of the propellants used in the APS module, leak checks had to be carried out with great care while wearing a protective suit. *(Courtesy of Phil Broad)*

the Saturn V that used *hypergolic* propellants, i.e. substances that merely have to come into contact with each other to begin combustion. Engines using hypergols were ubiquitous on the Apollo spacecraft, being used for the main propulsion systems as well as the attitude control thrusters. The fuel in the APS was monomethyl hydrazine ($CH_3 NHNH_2$) and the oxidiser was nitrogen tetroxide (N_2O_4).

To keep the propellant feed system simple and highly reliable, no pumps were used to force the liquids to the engines; they were fed merely by maintaining the tanks at a high internal pressure. The propellants were stored in two long tanks on either side of a module. Between them was a tank of helium gas filled to an extremely high pressure. This gas was fed via a control module, filters and check (one-way) valves to the tanks. As long as the feed pressure was greater than the pressure of combustion, this would be sufficient to ensure a supply of propellant on each firing.

An arrangement had to be made within the tanks to permit operation in a weightless environment. The task was to ensure that only propellant reached the tank's outlet, not the pressurising gas that occupied its empty volume. The method used in the APS modules (and on the Apollo spacecraft) was to separate the liquid and gas within the tanks with a bladder made from Teflon.

Thanks to its highly unreactive properties, Teflon is well known as a non-stick coating on pots and pans. The APS tanks contained a bladder made from Teflon that held the corrosive propellants and would not react with them. By pressurising the space between the wall of the tank and the bladder, the helium squeezed only liquid from the bladder towards the engine valves.

These little rocket engines burned merely by opening the electrically operated propellant valves to admit the two liquids into the combustion chambers. The small ullage motors had two valves, one each for fuel and oxidiser and they would fire continuously for a period of about 16 seconds when required.

Each of the three larger attitude control motors had four valves, two each for the fuel and oxidiser. They operated in parallel for greater redundancy whereby a single valve failure could be tolerated. These motors operated in short bursts of 70 milliseconds in order to maintain the stage's attitude to within a degree of the ideal.

Retro and ullage motors

At the end of the S-II's burn when the stage was nearly exhausted, it was jettisoned along with the conical interstage that had been built as part of the S-IVB. Tension ties that had held the structures together were cut by an explosive strip, separating the top of the interstage from the bottom of the S-IVB's aft skirt. A few milliseconds later, four forward-facing retrorockets built into the bottom of the interstage structure ignited to retard the spent stage's forward velocity as the S-IVB pulled free.

Each retrorocket consisted of a steel case, 2.3m × 23cm, filled with solid propellant that had a star-shaped cavity running its length. Such cavities are common in solid-propellant motors because they present a larger surface area for combustion at the start of the burn. This evens out the thrust profile of the rocket.

Each retrorocket was mounted slightly canted out from the centreline by 3° and the exhaust gases exited a nozzle that was

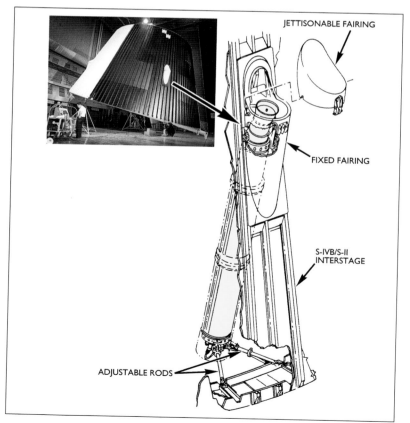

BELOW Cutaway drawing of a solid-fuelled retrorocket mounted in the S-II/S-IVB interstage. *(NASA/Woods)*

canted about 9° from the vehicle's centreline. Pyrogen igniters, essentially small sources of flame, initiated the combustion which provided 155kN (34,810lb-f) thrust for 1.52 seconds. A jettisonable fairing protected the upward-facing nozzle from contamination until firing.

As the S-II engines shut down, the rocket's structure unloaded in an instant, rebounding just like a spring. As the vehicle suddenly became weightless, the S-IVB's propellants were thrown forward for a moment, raising the possibility that ullage gas might come near the outlets of the tanks. This tendency was countered by two solid-fuelled ullage motors mounted to the outside of the aft skirt that were fired a fraction of a second before the stage separated from the S-II. Their acceleration forced the liquid contents to the bottom of the tanks, and they helped to pull the stage clear of the dead hulk below, cleanly extracting the J-2 engine from the interstage cone.

Each ullage motor held 26.7kg of solid propellant with a star-shaped cavity in a steel case, 87 × 18.5cm. They were ignited by pyrotechnic initiators to produce 15.1kN (3,390lb-f) thrust for about 4 seconds, aimed at an angle of 30° from the vehicle's centreline. About 12 seconds after ignition, having completed their task and with the J-2 now operating, they were jettisoned to further lighten the stage.

These ullage motors for the staging process should not be confused with the small ullage engines in the APS modules, which were too puny for this task and were used later in the mission.

Propellant dispersion

In common with the other Saturn V stages, the S-IVB had explosive charges affixed to the walls of the tank assembly in order to quickly disperse their contents into the atmosphere. This was to avoid a fully laden vehicle impacting the ground in the case of a major malfunction.

Were such a calamity to occur, the Range Safety Officer at the Launch Control Center would send a coded signal to a dedicated receiver that would detonate the explosives of the propellant dispersion system. The result would be to create a hole of 60cm radius in the bottom of the LOX tank and two parallel slits in the LH_2 tank, each of 6.15m length. The system was disabled at orbit insertion.

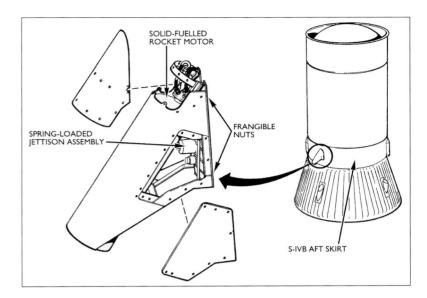

ABOVE Drawing of an S-IVB solid-fuelled ullage motor. *(NASA/Douglas Aircraft Company/Mike Jetzer/Woods)*

The S-IVB's legacy

Many of the S-IVBs eventually re-entered Earth's atmosphere to burn up, having been used as the second stages of a Saturn IB vehicle and thereby left in an orbit that would gradually decay. A few other examples that never got to fly populate various museums as part of displays of the Saturn IB and Saturn V, relics of the heroic age of manned space flight.

Five S-IVBs impacted the Moon in the name of science after having had an extra battery installed to support long-range tracking. A further five were sent away from the Earth/Moon system and into an orbit around the Sun. One of those, S-IVB-507 from the Apollo 12 mission, is believed to have been temporarily recaptured by the Earth/Moon system. When it was (re)discovered in September 2002, it was initially presumed to be an asteroid that had strayed into a chaotic Earth orbit, whereupon it was designated J002E3. But spectroscopic analysis indicated that its surface was rich in titanium oxide – the pigment of white paint.

Two other S-IVBs took a different path. They were converted into an entirely different type of spacecraft; a space station called Skylab. S-IVB-515 was a backup and is now in the Smithsonian National Air and Space Museum in Washington DC. The flight version, S-IVB-212, was launched on 14 May 1973 and stayed in orbit for just over six years. Its story will be the subject of the final chapter.

Chapter Seven

Instrument unit

From American rocket pioneer Robert Goddard's 1930s experiments through the A-4/V-2 wartime missile and the civilian Saturn V, advanced rockets of all types have required some form of control system to stabilise their flight and to guide them to their destination. This was the case whether the vehicle was carrying a weapon towards a target or an Apollo spacecraft to the Moon. The Saturn V relied on the instrument unit or just 'IU', a ring festooned with guidance equipment mounted above the S-IVB; the brains of the rocket.

OPPOSITE Near twilight, four days prior to launch, Apollo 8 atop its Saturn V vehicle stands on Pad 39A at the Kennedy Space Center on 17 December 1968. The pad is aligned to true north, orientating the rocket so that its pitch rotation will send it east to take advantage of Earth's rotation. *(NASA)*

ABOVE On 19 April 1932, Robert Goddard tested a rocket that included this gyroscope as part of its control system. *(NASA archives)*

BELOW An ST-80 inertial guidance platform as used in the Redstone missile. *(Mike Jetzer-heroicrelics.org)*

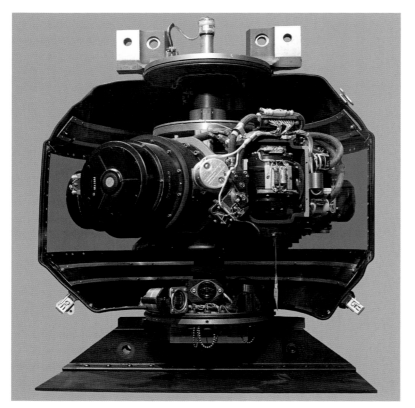

Central to the IU was the use of the gyroscope to provide a stable attitude reference. Goddard had suggested the gyroscope as a means to stabilise aircraft as early as 1907 and he went on to lay the foundations in the field and exploited them in his experiments on rocketry. His first gyroscopically stabilised rocket flew on 19 April 1932.

Goddard's achievements were appreciated by Wernher von Braun's team in Germany who applied them to the A-4's guidance system. Upon relocating to the United States, the team contributed directly to US capability in missile guidance in parallel with efforts by the US Navy and Air Force. Inertial systems developed rapidly for the multiplicity of rocket designs intended for nuclear missiles that came out of the 1950s and early 1960s as well as for submarine guidance.

The US Army team, centred around von Braun's engineers, created the ST-80 and ST-90 guidance systems for the Redstone and Jupiter missiles, respectively. When this team was incorporated into the Marshall Space Flight Center they combined the platform of the ST-90 with the ASC-15 flight computer that had been developed by IBM for the Air Force's Titan II missile in order to produce a guidance system for the early examples of the Saturn I vehicle. This started an association between the company and Marshall that would be maintained throughout the life of the Saturn programme.

Good as these early guidance platforms were, they came from an immature technological base at a time when huge strides were being made in the state of the art. Their limited operational life was more suited to a ballistic missile or a launch vehicle on a flight lasting only a few minutes. In view of the intention to have a Saturn operate in space for extended periods, MSFC set about developing a guidance platform that would be up to the task, the ST-124. During the latter part of the Saturn I programme, ST-124 units were flown along with the ST-90 to test their capabilities within early IUs that were mounted atop the S-IV upper stages.

In February 1964, IBM was named as the prime contractor for the IU to be used on the Saturn IB and Saturn V vehicles. This would be a modular design based around a 6.6m ring, with a hatch to allow easy access to

ABOVE The instrument unit for the AS-501 vehicle (Apollo 4) being hoisted within the VAB. *(NASA-MSFC)*

the equipment bolted around the inside. IBM created a facility near MSFC and set about working closely with NASA on the IU as the company took increasing responsibility for the gear installed within.

Structure

The skin of the IU had a construction style that was more in keeping with the Apollo spacecraft than the rest of the Saturn V. It was a sandwich of aluminium honeycomb faced by 7075-T6 aluminium alloy sheeting that was 0.51mm thick on the inside and 0.76mm thick on the outside. Including the honeycomb core, the walls were 24mm thick. These were produced as three parts of a ring, each a 120° arc, fastened together by splicing plates bolted across the joins to form a unit 6.6m in diameter and 0.91m tall.

Equipment was then mounted either directly to the wall of the IU or, if it required active cooling, onto one of the 16 cold plates that were fastened around the inside. The IU was subdivided into 24 equal sections for the purpose of locating equipment, and these were counted counter-clockwise from the splice nearest position 'I' of the rocket (facing east on the launch pad). The access hatch was on location 8.

A bracket was attached to the wall between each location to support a cable tray that ran all

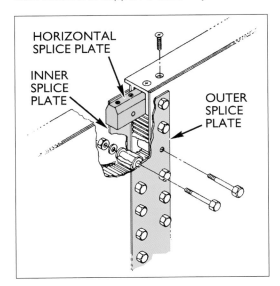

LEFT Details of the splice join that fastened the sections of the IU together. *(NASA-MSFC/Woods)*

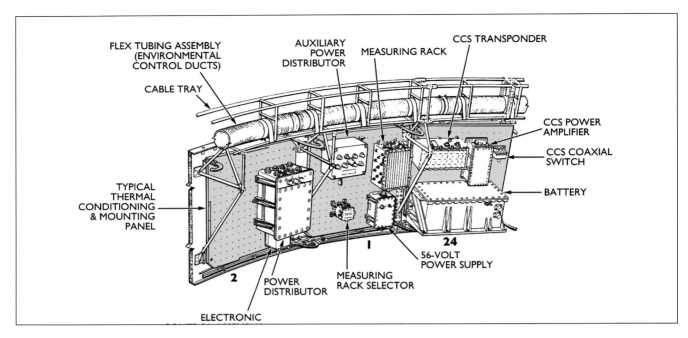
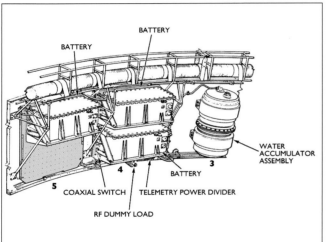
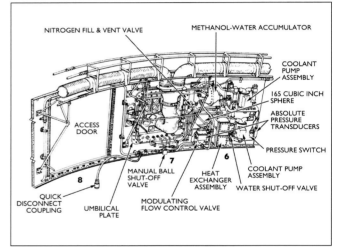
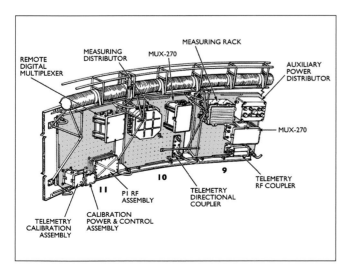
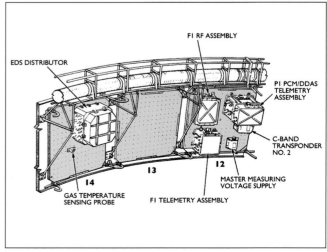

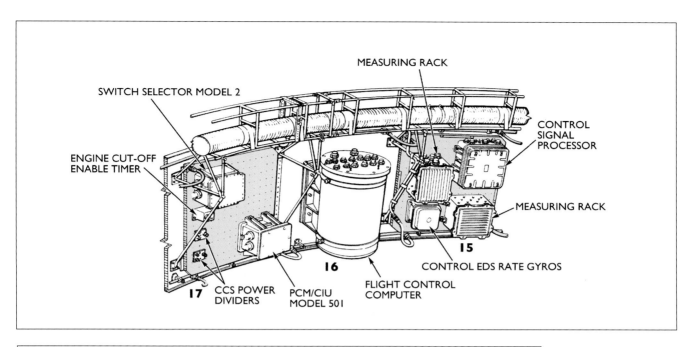

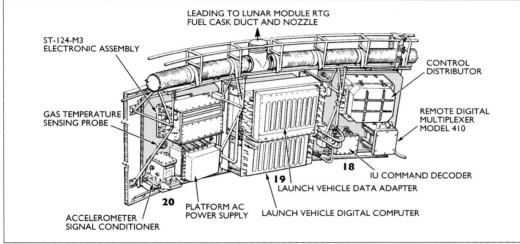

THIS SPREAD Layout of the eight sections of the AS-510 (Apollo 15) instrument unit. The cold plates upon which some equipment was bolted are indicated in blue. *(NASA/Woods)*

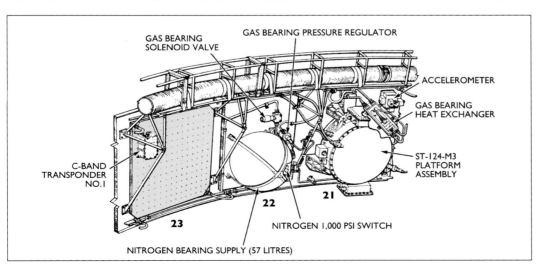

121
INSTRUMENT UNIT

around the IU above where the equipment was mounted. The tray had two channels; one for a flexible duct that was part of the environmental control system, the other to provide a conduit and support for the cable harnesses that serviced each module below.

Electrical power

Power for the IU's systems came from four silver-zinc batteries mounted at positions 4, 5 and 24, each delivering 28V. Although it was expensive, this battery technology was common in NASA's manned spacecraft because it was very efficient and tolerant of temperature extremes.

The 28V power was fed to redundant buses and extra feeds were used to generate a 56V supply for the guidance platform and a highly accurate 5.000V (±0.005V) used for sensors around the vehicle for precise measurements. An inverter generated a feed of AC voltage for those items that required it.

Environmental control

The IU used a thermal conditioning system which, during flight, used the same principle as that on the Apollo lunar module, namely the process of *sublimation*. When water is exposed to a vacuum, it vaporises rapidly and in so doing, it quickly loses heat. As it falls in temperature, it freezes and from then on passes from the solid phase directly to vapour, or sublimes, without going through the liquid phase; all the while taking heat with it.

In the IU, a heat exchanger, known as a sublimator, had purified water pass one side of a porous metal plate, the other side of which was exposed to the vacuum that existed inside the IU ring once the vehicle was above the atmosphere. As the water exited the pores of the plate, it froze to form ice that continued to sublime, cooling the plate for as long as a supply was maintained.

Multiple plates were arranged in pairs with coolant fluid running through passages between them, transferring its heat to the plates. Gaps on either side of each sandwich were exposed to vacuum. This heat exchanger conducted warmth from the coolant to the water and the porous plate, then on, through the ice, to space.

For the early IUs the coolant used was a mixture of water and methanol, but this was changed to a silicate/ester-based fluid. Having given up its excess heat in the sublimator, the coolant was forced around two circuits by redundant electric pumps. One circuit consisted of a set of cold plates in the S-IVB stage, just below the IU, which cooled its electronic equipment. The other circuit stayed in the IU, passing through its cold plates, the guidance platform, and the computer systems mounted around the inside of the ring.

The sublimator could only work in a vacuum so, prior to flight, cooling for the IU came from the ground via a second heat exchanger. For the 3-minute gap between lift-off, when ground thermal control was lost, until a time when there was sufficient vacuum within the IU for sublimation to work, there was no active IU cooling. However, it was felt that the thermal inertia in the system would keep the components within limits until the vehicle was above the atmosphere.

The story is told that when an engineer

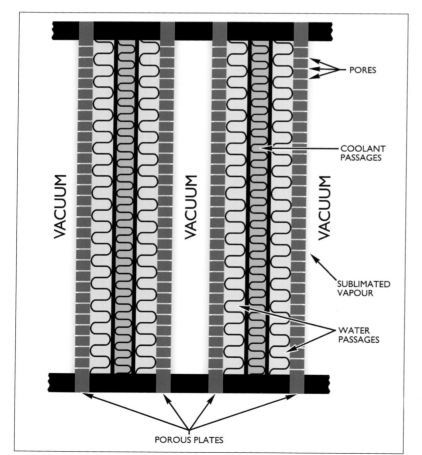

BELOW Diagram to show the arrangement of the IU sublimator. *(Woods)*

GUIDANCE, NAVIGATION AND CONTROL

When a space vehicle like the Apollo-Saturn V lifted off from Earth, it was trying to do two things. It wanted to be at a particular height at a specific time, and when it got there it wanted to be travelling in the right direction at a specific velocity.

Take Apollo 15 for example. By the time 11 minutes and 39 seconds had passed since lift-off, the stack was to be at a height of 171.5km (92.6 nautical miles) and travelling at a speed of 7,803m/sec (25,599ft/sec). Having flown around Earth's curve by some distance, the direction of its flight path at this moment should be 15° pointed down with respect to what had been the horizontal plane of the launch pad at the time of lift-off (345°, you can see that these things had to be carefully defined). By achieving those conditions, the stack would be in a nearly circular orbit at an altitude of 171.5km.

In order to get to that point, the IU needed to know the place from which it started and it needed to measure its acceleration as it headed to space. Why acceleration? If you know how you accelerate, then you know how your velocity has changed because acceleration is a measure of your change in velocity.

Likewise, if you know your velocity, then you know how your position has changed because velocity is a measure of how your position is changing. Therefore, if you know your starting position, then by keeping careful track of your acceleration, you can know your position at any point in your travel.

As an aside, this way of thinking about the rate of change of something lies at the heart of the mathematical technique called calculus. Separately invented by Leibnitz and Newton in the 17th century, it is an important part of so-called rocket science, yet the concepts it embodies can be intuitively grasped.

The IU not only measured its flight path, it could also control where it was going because it could swivel some of the engines by a few degrees to permit the rocket's thrust to be aimed. Since the IU knew where it was, where it wanted to go, and how to get there, it could send commands to ensure the vehicle followed the appropriate path to its destination.

ABOVE Page L2-3 from the Apollo 15 launch checklist gave expected values for every 30 seconds during ascent: angle with respect to the launch pad horizontal, inertial velocity and vertical speed (feet per second) and height (nautical miles). *(NASA)*

BELOW Instrument units being manufactured at IBM's facility in Huntsville, Alabama. The unit in the foreground, S-IU-504, would fly on Apollo 9. Behind it is the IU for Apollo 8. Both IUs are still in orbit around the Sun. *(NASA-MSFC)*

ABOVE The AS-501 (Apollo 4) IU is installed on top of S-IVB-501 within the VAB. *(NASA-MSFC)*

queried how much power the IU's equipment would have to lose, his colleagues gave somewhat conservative answers for each of their items. When the values were totalled, an extra degree of engineering conservatism was then added. Consequently, on the first flight of the IU, the sublimator froze up completely, there being too little heat in the system to maintain its proper operation. As a result, heaters had to be added in order to keep the coolant from icing up!

An additional task for the thermal control system was to cool the nitrogen gas used in the bearings of the gyros, accelerometers and pendulums of the guidance platform. A dedicated heat exchanger was installed for this purpose.

Before flight, the interior of the IU was filled with either air or nitrogen fed from a ground supply that was temperature-controlled and filtered. Air was used throughout the time leading to fuelling. Prior to starting to load propellant into the huge liquid hydrogen tank below, the atmosphere of the IU was replaced with nitrogen. The gas entered at the umbilical plate and was then routed through a flexible duct that ran around either side of the IU's walls. Holes in the duct allowed air or nitrogen to flow out over the equipment in the IU.

The remaining gas exited upwards through a duct 180° around from the inlet. This took it into the compartment where the lunar module was housed, where further ducting directed it to flow over a graphite cask on the side of that spacecraft's lower stage which held the fuel element of plutonium intended for a radioisotope thermoelectric generator that would power experiments to be left on the Moon. The fuel element was hot, giving off 1.5kW of thermal power and the purpose of the airflow was to conduct that heat away.

ST-124 guidance platform

At the heart of the IU's guidance and control system was the ST-124, a gyroscopically stabilised inertial platform built for IBM by the Bendix Corporation. A beryllium base formed the platform itself. To this were attached three gyroscopes and three accelerometers along with other items that helped with the alignment of the unit before launch. Three accelerometers were necessary to detect motion in all three axes of a Cartesian coordinate system, and it was important that they remained aligned with that system as the rocket rotated. This was the job of the gyroscopes.

If you hold a bicycle wheel by its axle and spin it up, you'll quickly discover that it strongly resists any attempt to rotate the axle away from its axis. It is this property that makes it easy to stay upright on a bicycle when it is moving. Miniaturise this idea whereby a small electric motor spins a disk at high speed. It is then possible to measure the force that the disk exerts on its mounts whenever an external influence tries to alter the alignment of the axle. That force can then be translated into an electrical signal. Three such gyroscopes installed orthogonally can precisely measure any rotation of the platform.

In the ST-124, the platform was mounted within a set of beryllium gimbals. Beryllium was used because it is one of the lightest metals yet is very strong, making it useful for space applications. Unfortunately, it is also quite toxic and requires great care when being worked.

The gimbals were a set of nested frames hinged at 90° intervals with respect to one another in order to give the arrangement three degrees of rotational freedom. The result was that as the outer support rotated by virtue of being attached to a vehicle that was itself rotating, the platform was free not to follow that rotation. It could therefore be actively driven to maintain a very precise orientation in inertial space; i.e. with respect to the stars.

Each gimbal was attached to its neighbour by two hinges placed 180° apart. One hinge included a small electric servomotor and the other had a *resolver*. The motor allowed the angle between the gimbals to be altered on command. The resolver allowed the actual angle between two gimbals to be measured very precisely. Thus, when a gyroscope detected a rotation, it could then signal that some correction was required. The appropriate servomotor could operate to counter that rotation and restore alignment. The complete assembly of platform and gimbals was sealed between two hemispheres to maintain control of the gas that was used within its bearings.

The upshot of this arrangement was that the platform kept a single orientation with respect to the stars as the rocket rotated around it. This stable platform was then a suitable surface from which the accelerometers could measure every motion of the vehicle and supply that information to its digital computer for processing. Additionally, the resolvers on

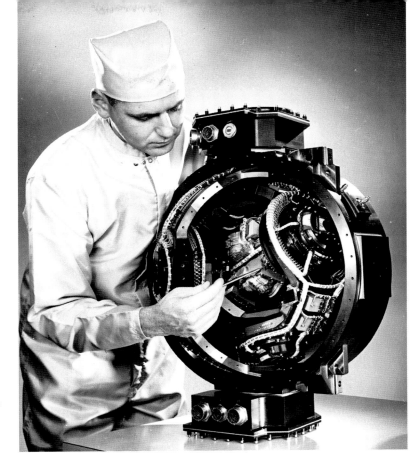

ABOVE An ST-124 inertial guidance platform without its covers. *(Bendix/NASA)*

BELOW Schematic diagram of the ST-124-M3 3-axis inertial platform. The adjoining image shows the relationship of the platform's axes with those of the rocket and the points of the compass. *(NASA/Woods)*

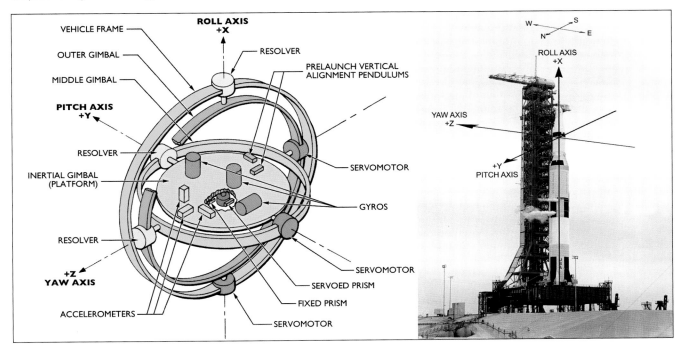

ABOVE Sensing its movements with an ST-124 and balancing itself on a column of flame, Apollo 15 rises from Pad 39A on 26 July 1971. *(NASA)*

the gimbals indicated to the computer which way the vehicle was orientated in inertial space, a key piece of information when directing the force of the Saturn V's enormously powerful rocket engines.

A rocket rising vertically off a launch pad is a bit like balancing a pencil on a finger. It was thanks to the ST-124 that a Saturn V, poised on its finger of flame, was able to rise gracefully from Pad 39 rather than topple over. It then steered itself along a carefully calculated flight path into the desired orbit around the planet less than 12 minutes later, coasting along at nearly 8km/sec.

Azimuth alignment

While sat on the pad, the vehicle was orientated in such a way that its coordinate system aligned with the points of the compass. Thus, the axis about which it pitched (its Y axis) pointed towards north. Since the ascent to orbit was intended to be a straightforward pitch manoeuvre from vertical to horizontal, if the vehicle were to launch straight up and then pitch over, it would begin its flight travelling directly east.

But no Apollo flight took a direct easterly path from the pad. They followed a particular heading, known as the *flight path azimuth*, that suited the goals of the individual mission. For the computer to be able to control the ascent in pitch alone the orientation of the ST-124's platform had to match this desired azimuth. That way the platform's Y axis would be perpendicular to the azimuth and its Z axis would be aligned along it. Then, upon lift-off, the whole vehicle would roll around its long (X) axis so that it too would be aligned with the desired azimuth, enabling the rest of the ascent to be a simple pitch manoeuvre that progressively tilted the rocket towards horizontal.

When aligning the platform to the flight path azimuth before a flight, a problem arose because of Earth's rotation. Whenever the newly aligned platform was allowed to maintain its own orientation, it would instantly begin to drift away from the proper alignment. In fact, the platform was doing its job correctly. It was the rotation of the planet that was inducing a drift rate of one complete revolution per day.

This situation called for the platform to be held in the required orientation with respect to Earth until a few seconds before launch. When that time came, Earth's rotation would have placed the launch pad in the desired orientation and the platform could be released to maintain its orientation with respect to the stars. This was the point, only 17 seconds before launch, when NASA's Public Affairs Officer would announce 'Guidance is Internal'. To the engineers it was formally known as *guidance reference release* (GRR).

To achieve an alignment, the system needed to know two directions. One direction was essentially downwards and this was determined by two pendulum devices on the platform, one of which sensed the true vertical along the platform's Y axis and the other did the same for the Z axis. Together, they indicated the true vertical. The second direction needed to achieve an alignment was to be along the flight azimuth. This was decided by a computer at the Launch Control Center, and depended on the exact time of launch. It was stated as a bearing with respect to true north, but its alignment could be determined with respect to any accurate reference. In this case, a theodolite was sited precisely due south and approximately 210m from the launch pad.

To accommodate this optical alignment technique, the casing of the ST-124 included a window which viewed through a 20cm aperture in the skin of the IU. A light from the theodolite was shone through the window. It was used by a servoloop as the reference against which to rotate the outer gimbal until the desired azimuth alignment had been achieved by the platform. This alignment was then held by the servoloop until it was released 17 seconds before launch. A retroreflector in the IU helped the theodolite maintain its aim even as the vehicle swayed in the wind.

This servoloop alignment arrangement was inherently flexible such that, were the launch to be delayed within its acceptable launch window, it merely required the launch control computer to calculate a revised flight azimuth and the platform would automatically rotate to align itself to match.

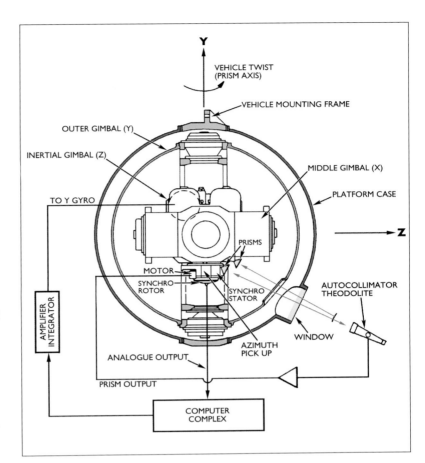

ABOVE Diagram of the servoloop control system that kept the ST-124 aligned using a theodolite until 17 seconds before launch. *(NASA/Woods)*

Digital computer

A significant part of the mythology of the Apollo programme is the incredulity people have about the power of the computer that landed men on the Moon. Consequently, a lot of attention has been given to this important machine, examples of which were built into both the command and service module and the lunar module of the Apollo spacecraft. It was thanks to NASA's relentless quest for ever more reliable units from the early microchip industry that not only were they able to shrink the size of the machine, in so doing, they primed the technology that has led to the modern computing era.

Missing from much of this early computer mythology is the fact that the Saturn V's IU also had a relatively small digital machine; the *launch vehicle digital computer* (LVDC). It was a true modern computer with random access memory (RAM) and the ability to run a stored program. But like the Apollo machine, it was more of an embedded controller rather than a general purpose computer in the sense we understand

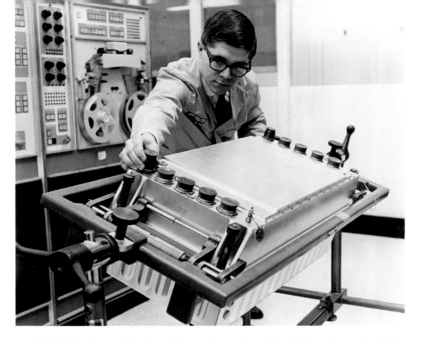

ABOVE The Launch Vehicle Digital Computer. With 32K words and a 2MHz CPU clock, it was an important development in small computers.
(Courtesy of IBM Archives)

BELOW Schematic diagram of the arrangement of the LVDC's ferrite core memory. Two intersecting wires flipped a core's magnetic field. The sense wire detected if a change had been made. *(Woods)*

today. It also used microchips as its logic building blocks and it too proved to be extremely reliable.

The tasks required of the LVDC were guidance and navigation, control of the Saturn V's attitude, control of events throughout the vehicle's life and general data handling, much of it to relating to the checkout of the vehicle, both on the ground and in orbit.

The computer's memory used an archaic technology in which small toroidal ferrite cores were magnetised. The direction of this magnetic field could then represent either a '1' or a '0'. To access and alter the state of the cores, wires were threaded through them in a grid pattern, the 'X' and 'Y' drive lines.

To read a single memory bit, DC current was applied to the two wires that ran through

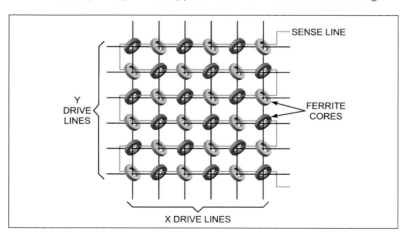

TIME BASE

A major task of the IU, and in particular of the LVDC, was to sequence the launch vehicle's various operations. For example, whenever falling propellant levels triggered the termination of the S-IC's burn, a complex sequence of events had to be initiated, each carefully timed with respect to the previous event; shutting down the engines, detonating the explosives to cut the stage free from the rest of the vehicle, firing the retro and ullage rockets (if present), starting the J-2 engines in the next stage and then jettisoning the interstage, etc.

As with this example, these sequences were always related to some important occurrence and were therefore programmed as a series of timed events set relative to that key moment. They were called *time bases* and a typical Apollo mission used eight major examples as well as a series of alternatives which hopefully would not be used.

Time base 1 – Lift-off

In order to ensure that this time base would not be triggered accidentally, it was inhibited by the LVDC unless 16 seconds had passed since the guidance platform had been released. This meant that the computer would only allow it to begin any time after one second to lift-off. It was then initiated by the disconnection of an umbilical as the Saturn V began to rise. Just in case that signal didn't get through, the LVDC was also programmed to monitor the *g*-force on the rocket soon after lift-off. If it sensed significant acceleration above 1*g*, then it started time base 1 anyway.

Time base 1 orchestrated the operation of the S-IC through most of its powered flight, separate to the guidance tasks that the LVDC was performing. Tasks included canting the four outer engines slightly so that, should one of them fail, the others were aimed more nearly through the vehicle's centre of mass to help to maintain control. It opened the valves that allowed helium stored within the LOX tank to be sent to pressurise the fuel tank. It issued a signal that would allow the Saturn's emergency system to shut down the engines if necessary after 30 seconds of flight (see page 131 for the explanation). Its final task was to shut down the centre F-1 engine after a preset time of about 135 seconds.

Time base 2 – S-IC cut-off coordination

The task of time base 2 was to coordinate the end of the S-IC's powered flight. Immediately after the centre engine had been shut down, the LVDC looked to see if the Saturn had achieved sufficient horizontal velocity. As soon as it had, time base 2 was initiated. This test was to ensure that launch had indeed occurred and that time base 1 wasn't inadvertently running on a stationary vehicle.

Its main task was to enable the systems designed to indicate that the S-IC's propellants were about to be depleted, and to arm the pyrotechnics that would perform staging.

Time base 3 – Staging and S-II control

Time base 3 was initiated by the same signal that shut down the S-IC's four outer engines, itself triggered by level sensors in the giant stage's tanks. It then orchestrated the dual-plane separation that jettisoned the S-IC stage and, 30 seconds later, the interstage ring.

Tasks included igniting the retro and ullage motors (if present) and coordinating the start of the S-II engines. It then took care of ancillary switching and telemetry calibration until, towards the end of the burn, it shut down the centre engine, armed the system that would shut down the outer engines, and armed the pyrotechnics that would perform the next staging event.

Time base 4 – Staging and S-IVB control

Similar to the previous time base, this one was initiated by the cut-off signal that also shut down the S-II's outer engines which, likewise, was triggered by the approaching depletion of the stage.

Time base 4 ignited the ullage rockets that would pull the S-IVB clear of the S-II and the interstage. It fired the pyrotechnics that separated the stages and commanded the stage's single J-2 engine to start. In addition to ancillary items, it commanded the solid-fuelled ullage rockets to be jettisoned once their job was done. Ten seconds after the J-2 engine had come up to power, time base 4 had completed its tasks. Thereafter it was up to the LVDC to monitor the vehicle's velocity and shut down the J-2 engine when it reached a precise value.

Time base 5 – Transition to orbital coast

Time base 5 began at the point when the S-IVB's J-2 engine was commanded to stop for the first time, this being the moment when the vehicle was inserted into a low Earth orbit. This time base then coordinated the safing of the vehicle for its period of orbital coasting and enabled the ullage motors in the APS modules to settle the propellants at the bottom of their tanks.

Time base 6 – S-IVB restart

Once the spacecraft's orbit had been accurately determined, computers at mission control could calculate the S-IVB's second burn, the translunar injection that would send Apollo to the Moon. These details were transmitted by radio to the IU, which took the scheduled time of ignition and back-timed by 9 minutes and 38 seconds. This became the start of time base 6 which would then orchestrate the steps leading to the TLI burn. This early start accommodated the time required to repressurise the propellant tanks.

Amongst the tasks performed within time base 6 was the opening of propellant lines to the O_2/H_2 burner and its subsequent ignition in order to warm helium stored in bottles within the LH_2 tank, expanding it to pressurise the tanks. Time base 6 closed valves to stop boil-off hydrogen being vented through the S-IVB's propulsive vents, and it preset the J-2's mixture ratio to 4.5:1, LOX to LH_2.

Time base 6 also prepared the J-2 for the TLI burn by starting pumps that circulated propellant through its systems, chilling it down so that only liquid would pass through the pumps without turning to gas. It then sent the command to start the engine.

Time base 7 – Begin translunar coast

Time base 7 began at the moment that the LVDC shut down the J-2 engine, which it did when it sensed that the required velocity for translunar injection had been attained. It then prepared the S-IVB for the coming operations. This sequence was mostly concerned with stabilising the stage and relieving the pressures in the tanks. This was necessary because the S-IVB had to be stable while the spacecraft extracted the lunar module from the top of the stage.

Time base 8 – Propulsive dump

Once the Apollo spacecraft had retrieved the LM and had backed away from the S-IVB, time base 8 was begun. It primarily dealt with the dumping of the remaining LOX through the engine in order to propel the stage onto a separate trajectory, usually to impact the Moon.

a particular core. The direction of this current would set the core's magnetic field to represent a zero. The grid layout meant that only one core would have current running through both wires that passed through it. Cores that had current from only a single wire running through them were not affected because the lower total current was not enough to alter their field.

If the core was already magnetised as a '0', then nothing happened. But if the core had been set as a '1', then its magnetic field would collapse and expand, this time in the opposite direction. The core now represented '0' where before it was '1'.

This change in the core's magnetic field induced a small electrical pulse in a third wire, the sense line, which had been threaded through all the cores in the grid. The important point to grasp is that the pulse would only occur if the bit had been flipped to '0', which meant it had previously been a '1'. The pulse then set a latch to indicate a '1', essentially transferring the contents of the bit from the core to the latch. This then allowed the computer to read it.

But by reading the state of the core, the read cycle had also altered it, so there had to be an additional store cycle in order to reset any '1' bits back to the '1' state. In all, the LVDC had access to 917,504 of these cores. This digital memory had to be carefully manufactured and assembled, often by hand!

One significant difference between the LVDC and the Apollo machine was that the latter had only 2 kilobytes of random access memory because the rest (36 kilobytes) was fixed, read-only memory in which to store programs. In the LVDC, all the memory was random access, including that used for program storage.

LVDC FACTS

CPU clock:	2.048MHz
Instruction cycle:	82.03 microseconds
Memory:	32K words
Word structure:	28 bits (26 + 2 parity bits) as 2 × 14-bit syllables
Instruction structure:	14 bits (4 opcode + 9 operand address + 1 parity bit)
Process:	Fixed decimal point, two's complement arithmetic

Emergency detection system

As a vehicle designed to carry humans to space, the Saturn V included systems to give the crew every chance to escape a major failure. Central to this was the *emergency detection system* (EDS) which gathered information from sensors across the vehicle to decide whether a dangerous situation was developing.

The EDS operated in two modes. If an emergency was developing quicker than the crew could react, it would intervene and unilaterally operate to save life. If the situation was developing more slowly, then the EDS merely presented the crew with the appropriate cues to enable them to work with mission control and reach a manual abort decision.

An important source of information for the EDS was a set of three gyroscopes that measured how fast the vehicle was rotating in roll, pitch and yaw. There were a further two sets of three, nine in all, to provide triple redundancy. Most of the ways that a Saturn V might fail would result in the vehicle's immediate loss of control and subsequent rotation at a speed much faster than would be expected during controlled flight. It was up to the gyroscopes to detect such rotation.

One of the most nerve-wracking moments of a Saturn V's ascent was during the flight of the S-IC stage when it was powering the vehicle through the thicker regions of the atmosphere. About 80 seconds into a flight, the vehicle passed a point known as 'Max Q' where the pressure caused by the airflow acting on an accelerating rocket (the aerodynamic pressure) reached a maximum before the rapidly thinning atmosphere reduced that pressure as it rose towards space. If control were lost around Max Q the situation would deteriorate very quickly, much faster than the crew could react.

The rules to initiate an automatic abort were threefold: 1. Structural failure between the spacecraft and the IU. (In other words, the conical spacecraft/lunar module adapter had failed.) 2. The thrust of at least two of the F-1 engines fell by at least 10% from their rated value. 3. Rotation of the vehicle exceeded 4° per second in either pitch or

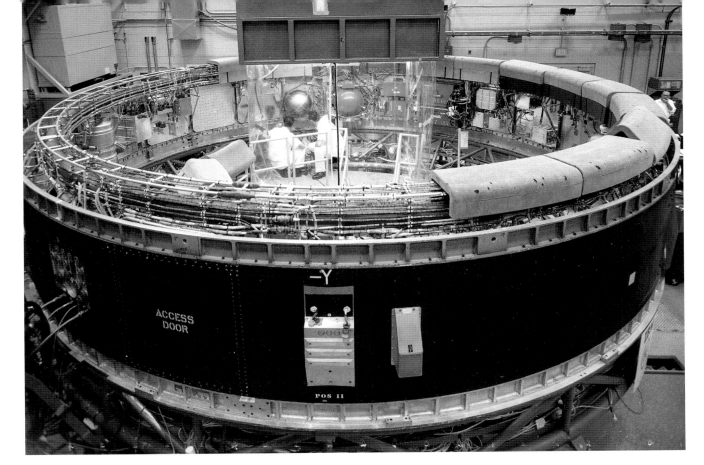

ABOVE An IU undergoing tests at IBM's facility in Huntsville. *(NASA-MSFC)*

yaw, or 20° per second in roll. In any of these situations, the EDS would immediately send a shutdown command to the F-1 engines while simultaneously cutting the Apollo command module free and firing the solid-fuelled escape rocket to pull it away from the Saturn.

The automatic mode of the EDS was switched off by the command module pilot two minutes into the flight. By that time, the aerodynamic forces on the vehicle were low enough that they would be unlikely to cause accelerated break-up of the vehicle. Abort decisions would then be taken by the crew in concert with mission control.

The ability of the EDS to command shutdown of the F-1 engines was inhibited by time base 1 for the first 30 seconds of flight. This was because it had been decided that as soon as the vehicle had lifted off the launch pad, even fractionally, there was no possibility of it being able to settle back down safely. Even with a failing vehicle, it was imperative that the engines continued to run to the best of their ability in order to gain as much height as possible, to allow the Range Safety Officer an opportunity to detonate the vehicle if necessary and disperse the propellant in the atmosphere.

When the Saturn V was conceived there was an expectation that it would be used to launch large advanced probes to the outer planets. It was for this reason that the vehicle was endowed with an independent guidance and control system, rather than depending on the perfectly capable G&N system within the spacecraft.

Nevertheless, from Apollo 11 onwards, the capability existed for an Apollo commander to wrest control of the Saturn from the IU and fly to orbit himself, based on cues from his displays. Indeed, some commanders dared the IU to quit on them just so they might get an opportunity to manually fly the mighty vehicle themselves, thereby further raising their pilot credentials. Not one of them ever got the chance.

The innate reliability of the IU and the wisdom of the decision to have the Saturn control itself were vindicated in spectacular style on 14 November 1969 when Apollo 12 was struck by lightning. (See page 142.) While the spacecraft's guidance systems were temporarily knocked out by the surge of electricity, the IU was blissfully unaffected and the Saturn continued to power to orbit under its own control as if nothing had happened.

Chapter Eight

The ascent of the Saturn V

Every journey of a Saturn V space vehicle began with sedate but impressive progress along the crawlerway at the Kennedy Space Center (KSC). With almost majestic poise, the complete but as yet unfuelled rocket, clamped to a heavyweight mobile launch platform and nurtured by a massive steel umbilical tower, was carried 5km to the base of an artificial concrete hill by a self-levelling, diesel-powered crawler/transporter.

OPPOSITE A crawler/transporter carries Apollo 14's space vehicle up the incline leading to Pad 39A on 9 November 1970. *(NASA)*
Top: Moments after Apollo 11 is released, its F-1 engines lift it from the launch platform. *(NASA/Stephen Slater)*
Middle: AS-503, the Apollo 8 Saturn V, powers away, tilting to the horizontal as it ascends. *(NASA/Stephen Slater)*
Bottom: Jettison of the Apollo 4 interstage. *(NASA/Stephen Slater)*

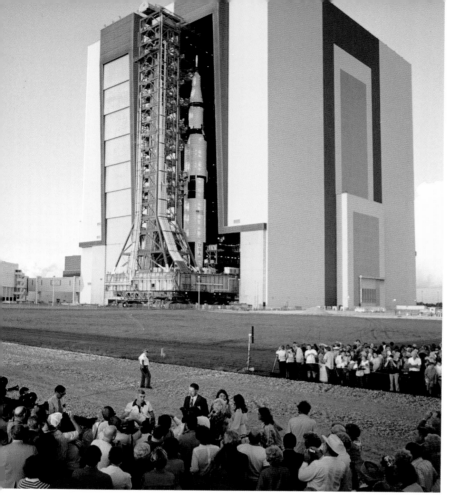

Then, with astonishing precision, the entire 8,400-metric-ton assemblage, still maintaining its level, ascended a 5° incline to where six strong pillars awaited to support the platform, tower and vehicle, ready for the next stage of its journey to space.

Smaller vehicles had traditionally been stacked directly onto the launch pad, but the sheer scale of the Saturn V combined with early intentions to fly large numbers of them in quick succession led KSC's planners to build a huge hangar, the *Vehicle Assembly Building* (VAB), in which the rocket stages could be stacked and tested while protected in an air-conditioned environment, away from the wind, rain and the corrosive salty air of Florida.

About three months before the scheduled launch date, with the Saturn V securely on the launch pad, a second tower was brought up to the vehicle to facilitate final servicing and checks. This *mobile service structure*, itself a 5,400-metric-ton creation in steel, stayed with the vehicle until the time came for propellants to be loaded, whereupon it returned to its parking site 2.4km back along the crawlerway.

ABOVE Apollo 17 commander Gene Cernan speaks to the crowd (foreground) on 28 August 1972 as the AS-512 space vehicle leaves the VAB mated to its mobile launcher and launch umbilical tower. *(NASA)*

RIGHT Apollo 16's space vehicle, AS-511, passes the parking site of the mobile service structure on 13 December 1971. Once the vehicle reached Pad 39A, the MSS was brought up to provide access and shelter to its upper sections. *(NASA)*

Mission	Apollo 4	Apollo 6	Apollo 8	Apollo 9	Apollo 10	Apollo 11	Apollo 12
Launch Vehicle	SA-501	SA-502	SA-503	SA-504	SA-505	SA-506	SA-507
Launch date (GMT)	09 Nov 1967	04 April 1968	21 Dec 1968	03 March 1969	18 May 1969	16 July 1969	14 Nov 1969
Launch time (GMT)	12:00:01	12:00:01	12:51:00	16:00:00	16:49:00	13:32:00	16:22:00
Launch mass (kg)	2,822,799	2,823,751	2,822,171	2,942,262	2,942,396	2,938,315	2,942,790
Flight azimuth (deg)	72	72	72.124	72	72.028	72.058	72.029

Mission	Apollo 13	Apollo 14	Apollo 15	Apollo 16	Apollo 17	Skylab 1
Launch Vehicle	SA-508	SA-509	SA-510	SA-511	SA-512	SA-513
Launch date (GMT)	11 April 1970	31 Jan 1971	26 July 1971	16 April 1972	07 Dec 1972	14 May 1973
Launch time (GMT)	19:13:00	21:03:02	13:34:00	17:54:00	05:33:00	17:30:00
Launch mass (kg)	2,949,136	2,950,867	2,945,817	2,965,241	2,961,860	2,852,959
Flight azimuth (deg)	72.043	75.558	80.088	72.034	91.503	40.880

ABOVE Launch statistics for all thirteen Saturn V flights. *(Woods)*

Alignment

At Launch Complex 39, both pads were built aligned to true north. This was entirely deliberate so that the Saturn V would be presented with its pitch axis also aligned north/south. Therefore, if the rocket needed to fly due east, it merely had to tilt around that axis, heading towards the east as it ascended. Due east was the most efficient direction in which to launch because it made full use of the direction and speed of Earth's rotation.

In the event, none of the Apollo flights headed in a precisely easterly direction. A bearing of 72° (about ENE) was the most common flight azimuth. The rocket therefore had to make a small roll manoeuvre as soon as it had cleared the umbilical tower to align itself with the desired direction. From that point on, all the major manoeuvring was a simple tilt motion in pitch. As a result, the crew would enter orbit in a heads-down, feet-up orientation – an attitude that also presented their spacecraft's optical systems away from Earth and facing the stars, ready for upcoming guidance calibration.

With the time of lift-off decided and the appropriate flight path azimuth determined, the guidance platform within the ST-124 unit could be aligned to match that azimuth. This was with respect to the theodolite sited directly south, 210m from the base of the vehicle. The platform would be held in that orientation with respect to Earth until being released a few seconds before launch.

Propellant loading

Between two and four weeks before launch, the lowest tank in the Saturn V was filled with over 800,000 litres of RP-1. This marked the start of a complete rehearsal of the events leading up to launch, including the loading of propellants. However, whereas the other tanks of the stack would be emptied of their cryogenics after the rehearsal, this great vessel of kerosene would remain full until launch because it was a stable liquid. It was finally topped off about an hour before lift-off.

The loading of the cryogenic tanks was somewhat more complex because they had to be purged of contaminants before their contents were introduced. (See the respective chapters for the details of propellant loading.)

Once the cryogenic propellants were aboard, their levels were maintained to compensate for the constant boil-off caused by heat leaking into the tanks. The S-IC required 1.3 million litres of LOX while the S-II and S-IVB required 331,000 litres and 92,350 litres respectively. The S-II required a million litres of LH_2 while the S-IVB took 253,200 litres.

Before launch, all the propellant tanks had to be pressurised. One reason for this was to present propellants to the inlets of the engine turbopumps at a high enough pressure to avoid the inducers and impellers being destroyed by cavitation. (See box on page 38.) Another was to provide additional strength to the tank walls, because these were important structural components of the Saturn V.

Some of the pressure at the pump inlets was achieved simply by the weight of liquid bearing down from above but a major source was the ullage pressure within the tanks, this referring to the space within the tank that was occupied by a gas rather than a liquid, with the pressure of this gas being transmitted through the liquid to the inlet of the engine.

S-IC pressurisation

Both S-IC tanks were pressurised by helium from a ground-based supply less than 2 minutes prior to launch. In flight, the fuel tank continued to be pressurised by helium supplied from four storage bottles carried within the LOX tank. This helium was piped to heat exchangers within each engine, where it was warmed by the hot turbine gas and expanded before being fed into the ullage space at the top of the fuel tank.

Shortly before launch, the vent valve for the LOX tank was closed and the continued boiling of the oxygen helped to raise the internal pressure to a desirable level. Helium gas completed its pressurisation until lift-off. As the LOX level fell, the tank's pressurisation came from gaseous oxygen generated by feeding LOX through the F-1 heat exchangers and back to the tank's ullage space.

S-II and S-IVB pressurisation

Prior to launch, all tanks for the upper stages were pressurised using ground-sourced helium. This provided pressure over that produced by the boil-off from the propellants. Then until it was the turn of that stage to perform, the boil-off maintained the tank pressures while relief valves protected the tanks against overpressurisation.

Again, like on the S-IC, LOX pressurisation in flight was provided by passing LOX through heat exchangers warmed by engine turbine gases to generate gaseous oxygen. This was then fed to the ullage space of the LOX tanks.

Pressurisation of the LH_2 tanks made use of the fact that the J-2 engine turned all of its hydrogen fuel into gas before being burned. This gasification occurred during its passage through the pipes of the thrust chamber. It therefore merely required a feed of this gas to be tapped off from the injector manifold on each engine and fed to the top of the fuel tanks.

Automatic sequence

All of the prepressurisation of the propellant tanks was handled by an automatic sequencer. This commenced at T minus 3min 7sec ('T' referring to the planned time of lift-off). Other important tasks carried out in these final moments included, at T minus 50sec, a transfer to the Saturn's own electrical power supplied by batteries on each stage and the IU. The internal hydraulic system for the F-1 engines was activated at T minus 30sec and the swing arms that serviced the S-IC were retracted.

T minus 17sec was the moment of guidance reference release, when the servoloop that had been holding the guidance platform aligned to the flight path azimuth let it go. The platform was now free to hold its orientation with respect to the stars. Defined relative to the horizontal plane of the launch pad at the time of launch and aimed along the flight path, this would provide its reference orientation in flight.

F-1 ignition

The S-IC's ignition sequence was begun at T minus 8.9sec. This gave enough time for all five F-1 engines to reach full power and for their stability to be verified by computer. But the commands to start the engines were not all sent simultaneously.

Once an individual engine had received its start command, it took about 6 seconds to work through its preliminary sequence. (See page 43 for a detailed description of F-1 ignition.) Then, as the engine's turbopump began to scream up to full speed, the thrust smartly ramped up towards its rated power

LEFT **Frames from film shot of Apollo 12's S-IC engines igniting. A pyrotechnic igniter within a nozzle can be seen near the top of the first two frames.** *(NASA/Stephen Slater)*

output. This transition to 6.7MN (1,552,000lb-f) took a further 1 to 1½ seconds. This 6sec delay was not consistent between engines and the precise time was determined during engine tests, especially tests of the complete S-IC cluster where start conditions more nearly matched those of launch.

Engineers preferred that the thrust build-up from all five F-1s should occur in a controlled fashion. Therefore to avoid an unpredictable application of the tremendous forces involved and to avoid all five engines applying full thrust at the same time, the timings of the start command to each engine were deliberately altered. What they wanted to achieve was a staggered start whereby the centre engine, no. 5, would ramp up first. Then, one quarter of a second later, diagonally opposite engines 1 and 3 should come up to power together. Finally, engines 2 and 4 should start together. It usually worked, but on a few occasions the timing for one engine was based on individual tests due to its late replacement. Nevertheless, no ill effects were noted from these out-of-sequence engine starts.

For about 1 second, maybe a little more, the 110m stack sat with its 3,000-metric-ton

BELOW **Graph of the transition of Apollo 11's first-stage engines to full thrust. The kink in the traces for the four outer engines was due to the ingestion of helium from the prevalves.** *(NASA/Woods)*

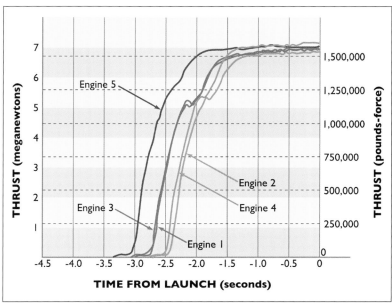

weight clamped to the pad by four hold-down arms. Below, five F-1s belched flame that exited at over 2.6km/sec while about 15 metric tons of propellant was consumed and the health of each engine was checked. A large wedge-shaped steel flame deflector below the vehicle diverted the exhaust horizontally north and south.

Lift-off

At the moment the decision was made to commit to flight, the clamps on the four hold-down arms were released simultaneously. As the vehicle began to rise, it pulled free of the controlled release mechanisms designed to impede the first 15cm of its ascent. Immediately, the umbilical carriers within three tail service masts alongside the base of the S-IC disconnected and rotated into their housings. Once in place, hoods came across to protect the fittings at the ends of the carriers from the blast. The hold-down arms were also protected by hoods pulled into place by lanyards attached to the rising rocket.

Of the nine swings arms that reached across from the launch umbilical tower, five had remained attached right up to the moment of launch to service either ends of the two upper stages and the service module of the Apollo spacecraft. With the vehicle now smartly rising, these arms had to disconnect their services and quickly rotate on hinges to place themselves snug against the tower's structure. To counter the momentum of tons of metalwork in motion and avoid crashing into the tower, aggressive braking was applied.

There was always some concern that the mechanism to actuate a swing arm would fail, leaving the appendage jutting out and in real danger of being struck by the rocket. To minimise this possibility, and to guard against a freak gust of wind pushing the Saturn V northwards towards the tower, the rocket performed an immediate yaw to the south, only straightening up again once it was clear of the tower, which took ten seconds. At this point, control of the mission switched from the Launch Control Center at KSC to the Mission Control Center in Houston, Texas.

Meantime, until 30 seconds into the flight, the emergency detection system was expressly forbidden to shut down the engines, even if they were failing. This was to provide time to fire the pyrotechnics in order to disperse the propellants in an abort.

At about 20 seconds, the outer engines were canted to aim their thrust away from the centreline of the vehicle. This was in case of a failure of one of these four engines. The

BELOW Frames from film footage of Apollo 12's lift-off show the moment that a hold-down arm (arrowed) retracts to allow the vehicle to rise. A tail service mast is in the foreground. *(NASA/Stephen Slater)*

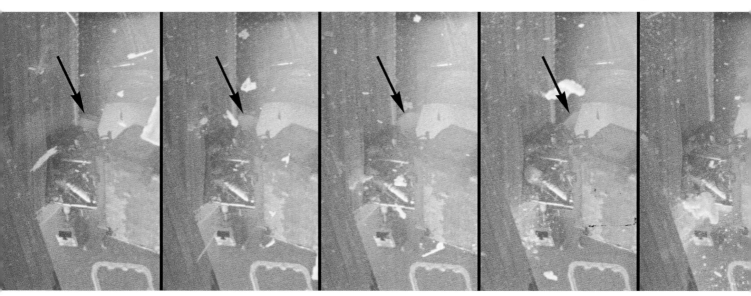

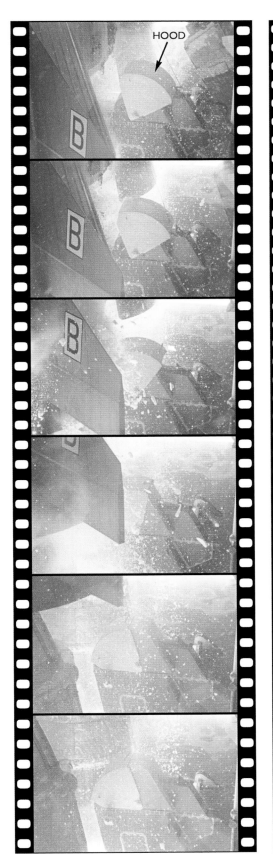

LEFT Retraction of a tail service mast is captured in these frames from film footage of lift-off. Once the mast is fully retracted, a hood closes to protect it from the blast. *(NASA/Stephen Slater)*

FAR LEFT As Apollo 12 began its ascent, a lanyard attached to the vehicle pulled a protective hood (arrowed) over each hold-down arm. *(NASA/Stephen Slater)*

ABOVE These still frames from film footage of Apollo 8's ascent show the retraction of the upper swing arms at the moment of lift-off. *(NASA/Stephen Slater)*

OPPOSITE A spectacular viewpoint at the top of the launch umbilical tower shows Apollo 14 as it begins its ascent on 31 January 1971. *(NASA)*

remaining overall thrust would act nearer to the vehicle's centre of mass, reducing the stresses on its structure and allowing more time to react to possible vehicle break-up, especially during the period of maximum dynamic pressure; Max Q.

Having straightened up, the next manoeuvre was to roll around to match the vehicle's coordinate system with the alignment of the guidance platform. This would also line it up with the flight path azimuth, typically 72°. For the rest of the S-IC's flight, the Saturn V would simply pitch over at a slow predefined rate.

RIGHT In this view from east of north, Apollo 8 can be seen to lean away from the launch umbilical tower in a yaw manoeuvre designed to help it clear any unforeseen obstructions. The crescent Moon has been added for artistic licence, as it was below the horizon at this time. *(NASA)*

ABOVE Apollo 8 gently tilts over in a pre-programmed manner designed to minimise sideways aerodynamic stresses. *(NASA/Stephen Slater)*

BELOW As Apollo 8's S-1C gained height, the reducing air pressure allowed its exhaust plume to greatly expand. *(NASA/Stephen Slater)*

Dumb or smart guidance

Ascent to orbit used two guidance modes. Throughout the S-IC flight, the Saturn flew in a 'dumb' mode. That is to say, it did not make any attempt to actively steer itself based on where it was and where it was going. Rather, it was flying a programmed 'tilt sequence' designed to minimise any sideways angle of attack as it headed up through the atmosphere.

The crew were able to monitor the angle of attack through the dual use of a meter in the spacecraft. Normally used to monitor the performance of their spacecraft's main engine, it was switched during ascent to display the output from eight pressure-sensing holes at the tip of the vehicle. In this mode, the meter was scaled so that 100 represented an abort condition.

The final act of the tilt sequence was to stop the vehicle from pitching over any further; known as *tilt-arrest*, this occurred shortly before the outer engines of the S-IC were shut down. During the process of staging, when the Saturn V was essentially cut in two, it was important to minimise any rotation that might cause the departing stage to recontact the remaining vehicle.

About 40 seconds after staging had occurred, with the S-II settled into its burn, the guidance went from dumb to 'smart' by initiating the *iterative guidance mode* (IGM). The vehicle was above the sensible atmosphere and could now safely steer itself to that desired point in the sky to insert itself into orbit at the correct speed and direction. The previous steering mode was open-loop. Any errors weren't compensated for but the guidance system nonetheless kept track of its progress. Now it was using closed-loop guidance which constantly compared the flight path with the ideal and undertook active steering to minimise the difference.

All-weather testing

The robustness of the Saturn V system was amply demonstrated when Apollo 12 launched in heavy weather on 14 November 1969. President Nixon was there to witness the ascent, perhaps adding a little extra pressure on the launch teams to get the vehicle off the ground on time. However, in the minds of those in the launch control building, there seemed little reason not to launch. Florida can be very humid and sees plenty of rainfall over a year. The Saturn V rocket often sat out in the rain.

On this occasion, although it was wet at the pad, the nearest lightning was many kilometres away and had not occurred at the site for six hours. The Kennedy Space Center wasn't in the middle of a thunderstorm. At 11:22 Eastern, the F-1 engines of the first stage lifted Apollo 12 on the start of its voyage to the Ocean of Storms on the Moon. Ironically, the stormiest moments of the mission were about to occur right here on Earth.

36.5 seconds into the mission, the stack was struck by lightning. In the spacecraft, the crew saw a flash of light and at the same time, the fuel cells that supplied electrical power were disconnected, forcing the spacecraft to switch onto its batteries. Immediately, the astronauts saw their caution and warning panel light up. As they took stock of their situation, the vehicle was struck yet again, this at 52 seconds into the flight.

The crew then had to come to terms with the fact that their guidance system was no longer aligned. The spacecraft had a gyroscopically stabilised guidance platform, very similar in operation to the ST-124 in the Saturn's instrument unit. The problem was that this platform was no longer stable.

"Okay, we just lost the platform, gang," yelled spacecraft commander Pete Conrad. "I don't know what happened here; we had everything in the world drop out." Conrad then began a litany of the caution and warning lights in front of him. "I got three fuel cell lights, an AC bus light; a fuel cell disconnect, AC bus overload 1 and 2, main bus A and B out."

Conrad's list was only the half of it. He couldn't see that in Houston, rows of engineers seated at their consoles could no longer see coherent streams of data coming from the spacecraft to inform them of its health. Fortunately, one controller, John Aaron, recognised the pattern of garbage and gave a call that has gone down in history.

"Apollo 12, Houston," called Gerald Carr, the astronaut tasked with speaking to the crew. "Try SCE to Aux." This switch was in an area of their huge instrument panel that was familiar to Alan Bean. He threw it, switching the spacecraft's signal conditioning equipment to its auxiliary power supply and thereby clearing the problem.

All the while, the first stage of the Saturn V below them continued to power the stack flawlessly, steered by the computer and associated equipment in the instrument unit. It was then cleanly shut down and cut adrift, enabling the S-II to ignite and continue the climb to orbit without missing a beat.

Meanwhile, Conrad had figured it out. "I'm not sure we didn't get hit by lightning," he opined. "Think we need to do a little more all-weather testing."

It is a testament to the robust design of the Saturn V that, despite the electrical calamity that had befallen its payload and the enormous electrical current that had flowed through the walls of the vehicle, the rocket had simply shrugged off the event like it never happened. Had the spacecraft been responsible for the

LEFT A still frame from film footage caught a discharge of lightning soon after Apollo 12's launch on 14 November 1969. *(NASA)*

Saturn's guidance, a task that it was expected to be capable of doing at a moment's notice, the rocket would surely have gone out of control and the mission immediately aborted.

This was a huge learning experience for the launch teams at KSC. What had not been realised was that because the gases of the superheated exhaust were ionised, the trail they left all the way to the ground made a very good conductor of electricity. Like Benjamin Franklin's experiment where he was reputed to have flown a kite in a thunderstorm to test the electrical nature of lightning, the rocket's exhaust acted as an enormous lightning conductor.

Though the raincloud's electric charge was insufficiently strong to cause lightning on its own, the rocket's exhaust essentially shorted the cloud to ground, thereby creating its own lightning. Then 15.5 seconds later, it did a similar trick, this time connecting one cloud to another to generate more lightning. Subsequently, NASA narrowed the criteria for acceptable weather at lift-off. They never again launched a Saturn V into a raincloud.

Max Q

The part of the ascent that held the greatest fear for the engineers was known as Max Q. When the vehicle sat on the pad, it experienced normal static sea-level air pressure which diminished to zero as it rose through a rapidly thinning atmosphere. However, as it accelerated, it was subjected to an additional pressure from its motion through the air, like the pressure felt when

ABORT MODES

When President Kennedy challenged his country to land a man on the Moon, he also stipulated that the fellow must be returned safely to Earth. This call for safety underscored the whole Apollo effort, and none more so than when negotiating the riskiest part of the journey; launch and ascent.

A great deal of time, effort and intense discussion by engineers distilled the options available to a crew in the event of a decision to abort a mission in its early moments. The possible scenarios were organised into a set of discrete abort modes which compartmentalised the ascent and specified the necessary actions, taking into account the changing conditions that would be met along the way.

There were four broad categories of abort mode, labelled I to IV. Mode I was further subdivided into A, B and C. The points in the ascent when they became relevant were announced by calls from mission control of "Mode One Alpha" and so on.

Mode IA
Applicable from when the crew were on the launch pad and valid until 42 seconds into the flight.
Like all Mode I aborts, the launch escape tower would ignite to haul the Apollo command module (CM) away from a failing Saturn V. After that, canards would be deployed (see Mode IB). However, because the vehicle would not yet have made any distance over the ocean, a small pitch control rocket motor in the side of the tower would fire to take the CM out over the water and away from any fireball that might be occurring below. From that point, the CM would jettison the tower, dump its manoeuvring propellant and deploy its parachutes for a water landing.

Mode IB
Applicable from 42 seconds up to an altitude of 30.5km (about 1min 53sec into the flight).
Now out over the ocean, the pitch control motor was no longer required. However, it was necessary to ensure that the CM with its tower was orientated properly with the heatshield first. During tests, the combination had been found to be aerodynamically stable in a pointy-end-first attitude at hypersonic speeds which would ram the tower and cover against the CM and make it impossible to jettison. Hence two canards were deployed at the top of the tower to force the CM around if necessary, and from there it would jettison the tower, dump its manoeuvring propellant and deploy its parachutes for a water landing.

Mode IC
Applicable from 30.5km to when the launch escape tower was jettisoned (about 3min 15sec into the flight).
This covered the end of the S-IC flight and the start of the S-II. An abort at this altitude was essentially above the atmosphere, rendering the canards ineffective, but the tower could be safely jettisoned and the CM's manoeuvring thrusters used to properly orient it for a normal water landing.

Mode II
Applicable from tower jettison to about 10 minutes into the flight.
Without an escape tower, the propulsive power of the Apollo service module (SM) would be used to pull clear of a wayward Saturn. Most likely, the small manoeuvring thrusters on that module would be fired to effect an escape but if necessary, the large engine would boost them away faster. Once clear, the CM would separate from the SM and make a normal landing.

For the remainder of the ascent, the abort options overlapped because there were choices about whether to use the S-IVB and/or the SM engine.

Mode III
Applicable from about 10 minutes until they reached orbit.
If required, Mode III would use the power of the SM to slow the spacecraft down in order to reach a predetermined landing site. The CM would then separate from the SM and make a normal landing.

Mode IV
Applicable from about 9 minutes until attaining orbit and also described as *contingency orbit insertion* (COI).
This was essentially an option to abort to orbit using the SM's large engine to make up the shortfall in velocity. The spacecraft would not get to the Moon, but instead would carry out an alternative Earth orbit mission.

S-IVB to COI
Applicable from about 6 minutes until the end of the S-II's burn.
This additional abort mode recognised that if a problem beset the S-II, then the S-IVB would be able to achieve a point from which the service module would be able to insert the spacecraft into orbit. Again, this contingency would close off the option of heading for the Moon.

holding a hand out of a moving car's window.

Known as Q, this dynamic pressure reached a peak, Max Q, about 80 seconds into the flight, soon after the vehicle went supersonic. Were things to go awry at this time, perhaps a control failure or loss of an engine, the Saturn V was likely to break up faster than the crew could react to abort the flight manually. This would possibly expose the spacecraft to a pressure wave from the explosive detonation of the LOX and LH_2 from the upper stages.

Riding the rocket

For the astronauts in the spacecraft at the top of the stack, the Saturn V was always a ride full of sensation and surprise, especially if they had never launched on one before. First stage flight was particularly energetic with lots of noise and vibration as the F-1 engines belched their fury below. Amongst all the shaking and rattling, the most impressive sensations were the abrupt side-to-side motions felt in the cabin as the rocket steered itself along its desired path.

The centre of mass of a fully loaded Saturn V was down in the first stage, thanks to the huge tanks of relatively dense RP-1 and LOX. The upper stages were comparatively lightweight owing to the low density of their hydrogen fuel. Consequently, small steering manoeuvres by the gimballed F-1 engines were amplified by the length of the structure on the opposite side of the centre of mass. Many astronauts agreed with Charlie Duke's sentiment on Apollo 16 that it was like riding a "runaway freight train on a crooked track, swaying from side to side".

As they sat on the pad, the astronauts experienced the 1g of Earth's gravity, but at the moment of lift-off this was augmented by the Saturn's acceleration. Two effects then occurred. At lift-off, the weight of the vehicle was only fractionally less than the thrust from the engines. This made for an initially gentle acceleration. But as it ascended, burning over 13 metric tons of propellant a second, the vehicle quickly lightened, making it more amenable to being accelerated. A secondary effect was that the efficiency of the engines increased as the atmosphere around them

LEFT Apollo 8's ascent midway through the S-IC powered flight. *(NASA/Stephen Slater)*

BELOW Graph of the acceleration felt by the Apollo 11 crew throughout their ascent. *(NASA/Woods)*

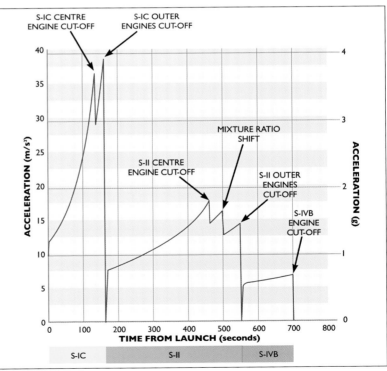

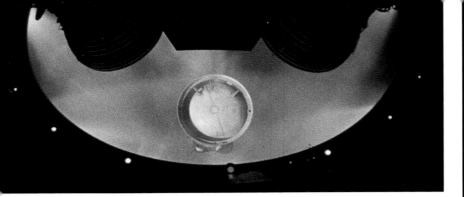

ABOVE Film cameras were carried aboard the early Saturn V flights to document the staging events. Here, Apollo 4's S-IC is jettisoned. The exhaust of the S-II ullage motors is visible as pink plumes.
(NASA/Stephen Slater)

RIGHT Thirty seconds after S-IC jettison, the Apollo 4 interstage was also jettisoned and fell through the S-II engine plumes as it did.
(NASA/Stephen Slater)

BELOW Graph of Apollo 4's five S-II engines transitioning to mainstage.
(NASA/Woods)

rapidly thinned, raising their thrust by 20%.

Many of the Apollo astronauts were already veterans of the Gemini programme and had ridden what had been designed as a nuclear weapon carrier, the Titan II vehicle. This converted ballistic missile had squeezed its passengers by up to 7g during ascent. The Saturn V, in comparison, was a gentle beast and its designers limited it to 4g. Nevertheless, by 2¼ minutes into the mission, the acceleration was on a steeply rising curve and heading rapidly towards this limit. To rein it in, the centre engine was shut down.

Suddenly, the acceleration dropped by a fifth but continued its rise towards 4g. Then, just as the limit was reached for a second time, all hell let loose.

Staging

When a 36-storey-tall tower constructed of relatively thin aluminium alloy is being accelerated at 4g, it will undergo significant compression. If the compression force is suddenly removed, like when four F-1 engines, each running at over 8MN, are suddenly cut off,

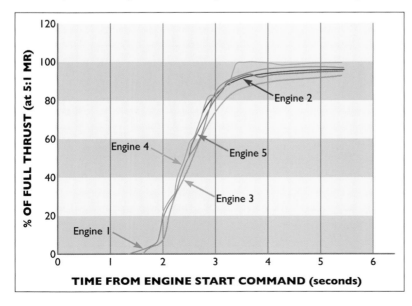

the structure will rebound rather violently.

"When that shut down, man, I thought I was going through the instrument panel!" Fred Haise had got the surprise of his life at the strength of the jolt when the first staging event occurred on Apollo 13. All at once, things happened quickly as the vehicle unloaded and various retro and ullage motors fired. Next, charges were detonated to cut loose the S-IC. Then the five J-2 engines on the S-II took over with an initial acceleration of just 0.75g. About 30 seconds later the interstage ring would also be jettisoned – except on one occasion when it didn't.

Skylab 1's interstage

Throughout the Apollo lunar flights, engineers believed that had the interstage failed to come away, an abort of the mission would be mandatory. They thought that within the confined space inside the ring with five powerful engines running, the heating would be too extreme. This assumption came to be tested on Skylab 1 and, fortunately, there were no humans riding the stack.

It was also fortunate that Skylab 1 was a lightly loaded Saturn V because it gave the stages the extra performance necessary to cope with serious failures; in this case a failure of Skylab itself.

Also known as the Saturn orbital workshop and constructed from a surplus S-IVB stage, it had a micrometeorite shield wrapped around its body. This was intended also to protect the workshop from the Sun's heat. Additionally, two solar panels were folded up and stored within arms which themselves would deploy after orbital insertion. These would provide the station with most of its electricity.

As the S-IC powered through Mach 1 and neared Max Q, the point of greatest aerodynamic pressure on the rocket, a design flaw at one end of a systems tunnel caused the micrometeoroid shield to peel away. As it did so it caused the unintentional release of a solar panel arm, which partially deployed. But the shield wasn't finished with the rocket. As it careered down the side of the second and first stages, it struck the explosive charge that would jettison the interstage from the S-II, cleanly cutting through it.

This charge was wrapped around the entire girth of the interstage and it was intended to sever 199 tension ties that held the ring to the S-II. As a result of the shield's impact, when the command was sent to jettison the interstage, the charge detonated only as far as the cut, leaving a substantial number of ties intact.

Still held by the remaining ties, the interstage sagged sufficiently to disconnect its electrical feed, thereby disabling an additional detonator that would have initiated the charge from the other direction. As a result the interstage stayed in place and engineers noted the rise in temperatures around the S-II's engines as a fraction of their exhaust gases played around the enclosed space. They elected not to abort the ascent.

An Apollo lunar mission would have been

ABOVE The AS-513 vehicle with the Skylab space station on Pad 39A. *(NASA/Ed Hengeveld)*

compromised by the dead weight of the interstage, 5.6 metric tons. But in this case the S-IVB was essentially an empty stage and there was no heavy Apollo payload above it. The S-II therefore had plenty of reserves to lift the workshop to orbit.

The interesting flight of Apollo 6

When the unmanned AS-502 vehicle launched as Apollo 6 on 4 April 1968, it was meant merely to reassure the engineers that the Saturn V's extraordinarily successful maiden flight as Apollo 4 had not been a fluke. It was fortunate that a second test of the vehicle was made, because much was learned from this flight, not least the excessive pogo vibrations two minutes into the S-IC's flight that would have severely shaken a crew.

Rather more serious was the faltering of the S-II, 5min 20sec into the flight. One of the outboard engines, number 2, showed a 3% drop in thrust. The engine ran on for another 1½ minutes until it and the adjacent outer engine, number 3, shut down together.

This really was serious. The S-II was designed to be able to handle the loss of an engine. It could even tolerate two as long as they were not adjacent outer engines. Yet, to the astonishment of the flight controllers in Houston, here was S-II-2 plodding on with two dead adjacent engines, finding enough ability to steer from the remaining engines so as not to tumble out of control. There was good reason to abort the flight at that point but instead, with no crew to worry about, they held back to see how the impaired rocket would cope.

Apollo 6 soldiered on and its three working engines used up the remaining propellant by burning for an extra minute. But the reduced off-axis thrust had taken the vehicle far from the programmed path and it was now up to the S-IVB to try to make its intended and actual paths converge.

This proved to be problematic for the single-engined stage, which began to snake across

LEFT Apollo 6 clears the tower on the Saturn V's second flight. *(NASA)*

the sky in an attempt to meet the intended conditions to enter orbit. It succeeded to a point and Apollo 6 was able to achieve a useful, if not precise, orbit around Earth.

But AS-502 didn't have its problems to seek because there was another J-2 failure. The S-IVB was to simulate an Apollo lunar mission by testing its ability to be relit. This would put the spacecraft on course for a high-speed re-entry to test the command module's heatshield. The restart failed and, instead, the service module's large engine had to be used for the task, or at least a lower-power version.

As if this litany of problems wasn't enough, the four-sectioned conical shroud atop the S-IVB that supported the spacecraft and which would house the lunar module had begun to shed pieces during the first stage flight.

The reason for the failing engine 2 on the S-II exposed a small but fatal flaw in the design of the J-2. (See chapter 4 for a detailed description of the J-2 engine.) At the centre of the injector was the *augmented spark igniter* (ASI) which was fed with LOX and LH_2 to generate a flame that ensured burning across the injector face.

The liquid hydrogen feed was taken from the LH_2 high pressure system at a point after the main fuel valve. It was then fed along a pipe to the ASI. In order to accommodate the inevitable expansion and contraction of a line that would carry supercold LH_2, two sections had been made flexible by adding bellows, themselves protected by a metal braid. However, when high-pressure fuel was fed along the pipe, the bellows resonated violently and failed through metal fatigue.

This small leak hardly affected the pressure of the fuel system but it reduced the feed of fuel to the ASI, causing it to burn LOX-rich. This created very high combustion temperatures that eroded the injector and LOX dome. Debris from the erosion damaged the tubing in the thrust chamber, further reducing the fuel pressure. Eventually, the LOX dome failed, causing the engine to be shut down.

Engine 2 indicated to the IU that it was in trouble and was about to shut down. The IU responded by closing the prevalves that led to that engine. Unfortunately, a wiring error caused the LOX prevalve leading to engine 3 to be inadvertently closed and this perfectly healthy

ABOVE Apollo 6 tilts over during S-IC boost. The vehicle experienced major pogo vibrations during this time. *(NASA)*

BELOW A test of a J-2 injector burning LOX-rich in the ASI caused major erosion of the structure. *(NASA/Woods)*

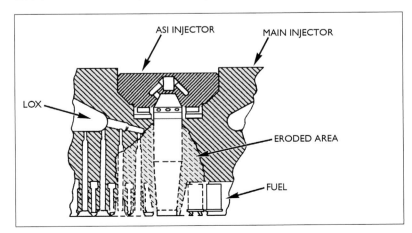

engine promptly shut itself down also.

So why had the design weakness with the bellows not been caught during the extensive ground tests on previous engines? It was discovered that when testing within the atmosphere, the extremely low temperature of the LH_2 passing through the bellows caused the air around them to liquefy. Thanks to the metal braiding, this liquid was trapped around the bellows and served to damp their tendency to resonate.

When the engine was operated in a vacuum, this liquid was absent, allowing the bellows to oscillate wildly until they failed, which they did on engine 2 of the S-II. Further analysis revealed this as the cause of the problem involving the J-2 on the S-IVB. There, the first burn was not long enough for the problem to reach the point of complete failure but the damage to the high-pressure fuel system was enough to prevent a restart.

Knowing what had caused the failure allowed engineers to address the problem. The LH_2 pipe that led to the ASI was redesigned to use bends rather than bellows as a means of accommodating the expansion and contraction.

The investigation of all the Apollo 6 anomalies was an engineering tour-de-force typical of NASA at that time. The J-2 problem was addressed, a fix was in hand to cure the S-IC pogo using helium in the LOX prevalves (see page 58) and the lunar module's shroud had been redesigned. The all-up mode of testing the Saturn V was an inspired decision that had proved itself. Having gained enough trust in the machine, Apollo's managers made their gutsiest call by agreeing that the next vehicle, AS-503, should not only carry three men, it should send them and their Apollo 8 spacecraft out to orbit the Moon.

S-II pogo and Apollo 13

Though there were always a series of minor niggles that affected every flight of the Saturn V, the only remaining major problem that continued to plague it was the incidence of pogo that arose on the S-II's centre engine towards the end of the stage's burn. It first reared its head on AS-503 and Apollo 8 commander Frank Borman, a tough, no nonsense military pilot who avoided hyperbole, reported, "Quite frankly, it concerned me for a while, and I was glad to see the S-II staging." That was damning from him.

Engineers noticed the same phenomenon on Apollo 9. The centre engine on S-II-4 had begun to oscillate at 16Hz, backwards and forward on its crossbeam support at a potentially damaging ±12g. Rather than making structural changes to the already feather-light stage, they opted to simply shut the centre engine down 1½ minutes early and complete the burn using just the four outer engines.

This appeared to have eliminated the problem. Indeed, in his autobiography, Apollo 11's Michael Collins described S-II-6 as "smooth as glass, as quiet and serene as any rocket ride can be." But on Apollo 12, a rash of pogo episodes came and went during the S-II burn, only completely clearing when the centre engine shut down. Since the vibrations were only one-third as strong as previously, it was decided not to change anything for the next flight, Apollo 13.

Vehicle AS-508 lofted the Apollo 13 spacecraft on 11 April 1970. As the S-II took up the task of boosting the spacecraft to orbit,

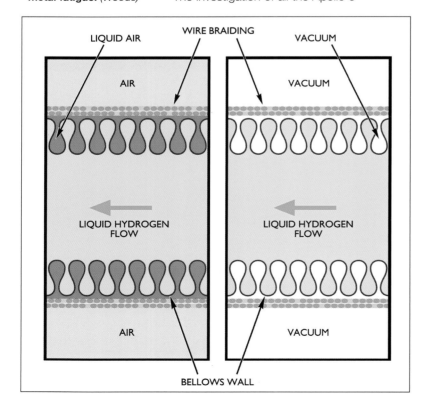

BELOW Schematic diagram of the ASI line's bellows section. In air, liquid air condensed around the bellows, dampening their vibration. In a vacuum, this was missing, leading to metal fatigue. *(Woods)*

it exhibited bouts of pogo similar to what had been experienced on Apollo 12. The third episode, however, became extremely violent with the centre engine, mounted in the middle of its crossbeams, undergoing back and forth shaking at 16Hz that peaked at ±33.7g, far beyond the loads that the beams were rated for. Then, 2min 12sec earlier than scheduled, this engine shut itself down.

What appeared to have happened was that the LOX tank pressure on Apollo 13's S-II was slightly low; only by about 20kPa (3psi) and still within set limits. During pogo, the longitudinal movement of the engine caused the inlets to the turbopumps to experience pressure variations. In this instance, the low part of the pressure cycle was low enough to cause cavitation at the LOX inlet to engine 5. (See box on page 38 for explanation of cavitation.) Unfortunately, the natural frequency for the creation and destruction of the cavitation bubbles was similar to the pogo frequency, amplifying the pressure variations. The LOX turbopump itself then began to cavitate, which affected the engine thrust in a way that reinforced the shaking.

In very short order, the outlet of the LOX turbopump showed such extremes of pressure that the 'Thrust OK' switches were tripped, shutting down the engine. The reduced overall thrust was somewhat compensated for by burning the remaining four engines for longer. Then the shortfall caused by a dead engine having to be carried for longer was made up by the reserves available in the S-IVB. As the S-IVB performed its second burn to put Apollo 13 on a lunar trajectory, Jim Lovell hoped his mission had just experienced its only serious glitch.

In the light of this continuing pogo problem, engineers added an accumulator, essentially a cavity, to the LOX duct that fed the centre engine on future S-II stages. This altered the duct's resonant frequency.

Mixture ratio change

As the S-II emptied its tanks, the acceleration felt by the crew increased from about 0.75g to 1.8g. When the centre engine was shut down towards the end of the burn to avoid its pogo tendency, the acceleration fell back to about 1.5g. Soon

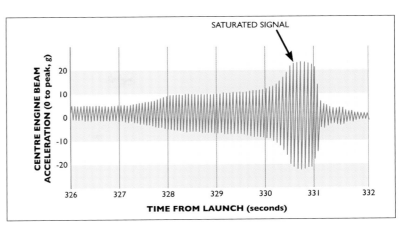

ABOVE Graph of vibrations measured at Apollo 13's S-II centre engine beam. At 331 seconds, the vibrations became so large that the sensor became 'saturated', unable to measure the ±33.7g shaking. *(NASA/Woods)*

after, the acceleration dropped again owing to the mixture ratio of the propellants in the engines being altered, though it was not always discerned by the crew.

Up to Apollo 13 the S-II stages had J-2 engines whose mixture ratio could be set to low (4.5), null (5.0) or high (5.5). From Apollo 14 onwards they had only low (4.8) and high (5.5) settings. Most of the burn was carried out with a high mixture ratio that yielded a high thrust but a relatively low efficiency. When a specified velocity had been reached, the MR changed to a low setting using less LOX than before. This lowered the thrust but increased the efficiency, and was intended to ensure that as little propellant as possible remained at the end of the S-II's burn.

The final act of the S-II's burn was to prepare for the upcoming staging by bringing any rotation of the stage to a halt, just as had been done at the end of the S-IC burn. In the iterative guidance mode that the vehicle was using, this was known as 'chi freeze', essentially freezing the flight path angle.

The S-IVB's first burn

As soon as the S-II's engines cut out, the vehicle began a sequence to jettison the empty stage and ignite the S-IVB's single J-2. Powerful ullage rockets on the S-IVB ignited just as charges were detonated to cut the S-II adrift along with the conical interstage. Retrorockets within the interstage were also ignited to help pull the two stages apart. The start sequence for this first burn of the J-2 included a three-second delay to allow more time for the thrust chamber to chill, compared to just one second

ABOVE **Still frame from film footage of AS-202's S-IVB staging. This 200-series stage had three solid ullage motors rather than two on a Saturn V, for which no footage of separation exists. The clockwise roll motors on the APS modules can be seen firing.** *(NASA/Stephen Slater)*

that the S-II engines had required.

Once the J-2's burn was established, the ullage motors were jettisoned and the computer cancelled its chi-freeze, allowing the smart mode of guidance to re-establish itself. The stage fired for about 2½ minutes to add the final 10% of velocity that would insert the Apollo spacecraft in the required orbit.

Translunar injection

The primary role of the S-IVB on an Apollo lunar mission was to be reignited and burn for about 5min 50sec in order to increase its velocity from 7.8km/sec to 11km/sec. This translunar injection (TLI) burn would place the spacecraft in a very long, elliptical orbit around Earth whose apogee would be half a million kilometres distant.

The timing and geometry of this trajectory was chosen so that Apollo would rendezvous with the Moon and pass its leading hemisphere in such a way that lunar gravity would sling the spacecraft straight back to Earth. This so-called 'free-return' gave an Apollo crew the option of either flying straight home or going into orbit around the Moon as they passed over the lunar far side.

Once the time for TLI had been calculated, the information was uplinked to the IU, which back-timed 9min 38sec to determine the start of time base 6. This would orchestrate the events prior to ignition. The activity began by closing vents and pressurising the propellant tanks by introducing warmed helium which had passed through the coils of the O_2/H_2 burner.

When a crew headed away from Earth, they would be utterly dependent on the machinery around them as a quick return to Earth would be impossible. They had to be completely confident in their systems and the diligence of the teams supporting them. It was a profound moment and not one to be approached lightly. They were therefore given an opportunity to instruct the IU that they were ready to continue by setting the XLUNAR switch. As part of its preparations for the TLI engine burn, the IU checked to see whether that switch had been placed in its INJECT or SAFE position, indicating whether preparations ought to continue.

With 1min 40sec to go before J-2 ignition, the LVDC began to aggregate the acceleration that the vehicle would undergo during the

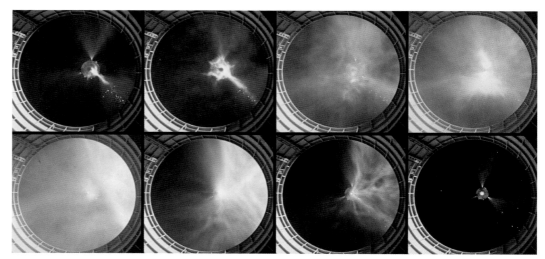

RIGHT **Sequence showing J-2 engine start for AS-202, filmed from the S-IB booster.** *(NASA/Stephen Slater)*

burn. At 1min 20sec the ullage motors in the two APS modules fired to settle the S-IVB's propellants. Eight seconds to ignition, the J-2's main fuel valve began to open, allowing cold fuel to circulate through the pipework of the thrust chamber. At the correct time, the contents of the J-2's start tank were discharged through the pump turbines, spinning them up and pushing propellant into the thrust chamber to be ignited by the flame from the ASI. The TLI burn had begun.

Throughout the burn, the IU kept the stack pointed forward, gently steering it as it came around the planet. Should the IU fail at this point, the commander had the option of steering manually.

In the early part of the burn, the crew felt a surge of power as the engine went to the high mixture ratio. To this point, it had been running fuel-rich (a ratio of 4.5:1) in order to use up the spare LH_2 that had been loaded in case an extra orbit of Earth had been required. It then moved to a ratio of 5.0 which raised the thrust by about 13% for the remainder of the burn.

As soon as the required velocity had been achieved, the IU shut down the J-2 and opened valves to vent the pressure from the tanks. The moment of translunar injection was defined to be ten seconds after shutdown to allow for the tail off in the thrust and the vehicle was now on its translunar coast. The tasks remaining for the S-IVB were few and limited.

Bound for the Moon

The final burn of the S-IVB had raised the stack's velocity to a little over 11km/sec and as a result, it began to climb away from Earth as it headed out on the start of its long, elliptical orbit. For a few minutes, the S-IVB aligned itself with the ground below as it hurtled around the planet, spacecraft forward but with the crew's heads down. Then it fired its APS thrusters to adopt an attitude suitable for the next part of the mission, the retrieval of the lunar module, nestled in front of the hydrogen tank's forward bulkhead.

This exercise was called *transposition, docking and extraction* (TD&E) and it required the S-IVB to adopt an orientation that allowed the crew to view the lander with oblique

lighting across its docking equipment and without the Sun in their eyes. As the APS thrusters held the S-IVB steady, the Apollo CSM separated and pulled away. The shroud that had enclosed the lunar module was cut into four pieces, each of which swung away and detached, revealing the lander.

As the CSM drifted away, it rotated to face the S-IVB. The drift was halted and it was then slowly brought up to dock with the lunar module. After connecting umbilicals within the docking tunnel and checking that all was well, the crew issued a command to fire the four pyrotechnic fastenings that held the lunar module to the third stage. The complete Apollo spacecraft was now free to make its own way to the Moon. But the IU was not yet finished with the S-IVB.

Lunar impact

Up to and including Apollo 12, flight controllers in Houston deliberately slowed the S-IVB in its ascent to the Moon so that by the time it reached lunar distance, the Moon would have gone past. Thus whereas the spacecraft would fly around the leading limb, the spent stage would coast past the trailing hemisphere and gain a gravitational slingshot effect that would accelerate it out of the Earth/Moon system into solar orbit.

The Apollo 12 crew were the first to deploy a sophisticated science station on the lunar surface. It included a sensitive seismometer for measuring moonquakes and the S-IVBs

ABOVE Lunar module *Challenger* mounted on top of Apollo 17's S-IVB on 7 December 1972. *(NASA)*

ABOVE A docking failure caused Apollo 14's CSM to remain with its S-IVB long enough to see its hydrogen fuel being released from non-propulsive vents. *(NASA)*

from subsequent missions provided a handy source of energy for generating the vibrations that would help characterise the internal structure of the Moon. Each S-IVB would represent a known mass travelling at a known speed and therefore provide a known energy upon impact. Its guidance system would allow the stage to be tracked all the way down and so the seismologists would know where the impact occurred.

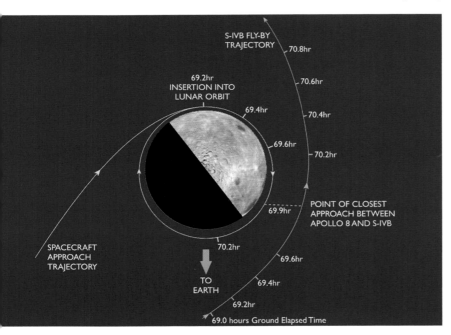

BELOW The trajectories of Apollo 8 and its S-IVB. While Apollo 8 entered lunar orbit, the rocket stage was given a gravitational slingshot into solar orbit. *(Woods)*

As the warmth of the Sun continued to heat the propellants of the spent stage, now flying alone, valves were opened and the tanks vented through orifices located 180° apart so as to cancel out any thrusting effects.

To hit a desired target on the Moon, the engineers needed to carefully utilise the few remaining sources of propulsion in order to retard the stage's velocity by just the right amount. They had whatever propellant was left in the APS modules and the thrust that could be obtained by dumping the stage's remaining propellant, particularly LOX. Other dumps would have to be made in a non-propulsive manner.

First, a burn was carried out by the rear-facing APS ullage engines in an evasive manoeuvre that put some distance between the S-IVB and the spacecraft. The stage then rotated to face the J-2 engine in the direction of travel in preparation for a dump of the LOX in a retrograde direction to slow the stage down.

By opening both the LOX prevalve and main valve, a substantial portion of the remaining LOX was allowed to exit through the J-2's engine bell under pressure from the tank itself. This was the main source of propulsion to aim at the Moon. In the case of Apollo 15, for instance, 1.17 metric tons of LOX was dumped for 48 seconds to achieve a thrust of 3.15kN (709lb-f). This changed the stage's velocity by 9.1m/sec (30ft/sec). The remaining LOX was dumped through the non-propulsive vents.

A small additional amount of propulsion could be gained by dumping the helium remaining in the stage's storage spheres but the final impulse came from burning the APS ullage motors almost to their depletion. The last act was to initiate a slow roll of the stage to even out temperatures across the stage and help cancel out any remaining minor thrust as gases continued to vent from its tanks. In the event, this roll soon became a tumble as the momentum of the rotation was transferred to a more stable axis.

The S-IVBs that were destined to hit the Moon were given an extra battery to enable the ground to maintain contact and track the trajectory right to the moment of impact three days later in order to determine the site with some accuracy.

Unfortunately, the radio system on Apollo

16's S-IVB failed 27 hours into the mission, creating significant uncertainty in its impact point after two further days. When the Lunar Reconnaissance Orbiter photographed the Moon in great detail almost 40 years later, the 30m craters made by most of the S-IVBs were easily located, but it took over six years to find the impact site of Apollo 16's S-IVB.

The impact of the S-IVB stages and of the discarded landers allowed the various seismometers across the near side of the Moon to be calibrated. The propagation of the shock waves through the surface showed that the Moon is a relatively solid body. Without a substantial molten subsurface, seismic waves could travel long distances before losing their energy. Geologists often explain the longevity of the signals as being as if the Moon was ringing like a bell. However, they did locate a major discontinuity in the crust at about 60km depth dividing the crustal material from the denser mantle.

The historic Saturn V

The Saturn V stages in various museums enable us to marvel at the scale and audaciousness of these extraordinary machines. Those that flew are either lying at the bottom of the Atlantic Ocean or were vaporised upon re-entry into Earth's atmosphere or upon impact with the lunar surface. A select few S-IVBs, including from Apollo 11, coast around the Sun in a similar orbit to Earth's, returning to the Earth/Moon system on occasion. Perhaps one day, a space-faring civilisation might choose to locate these venerable stages and return them to the planet of their birth, icons of a golden age of exploration by ancestors whose dreams were boundless.

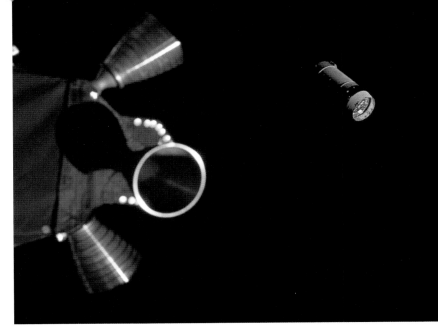

ABOVE Apollo 14's S-IVB-509, soon after lunar module *Antares* had been pulled free. This stage coasted to an impact within Mare Cognitum. *(NASA)*

BELOW Five S-IVB stages, all shown in this montage, were steered to impact the lunar surface to produce calibrated seismic events that probed the Moon's interior. In the process, each impact formed a new 30m crater. *(NASA/Goddard/Arizona State University/Woods)*

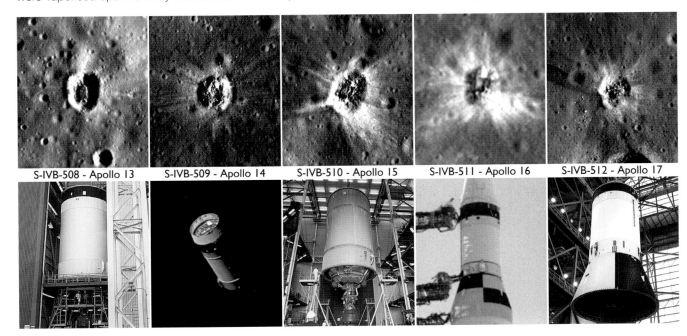

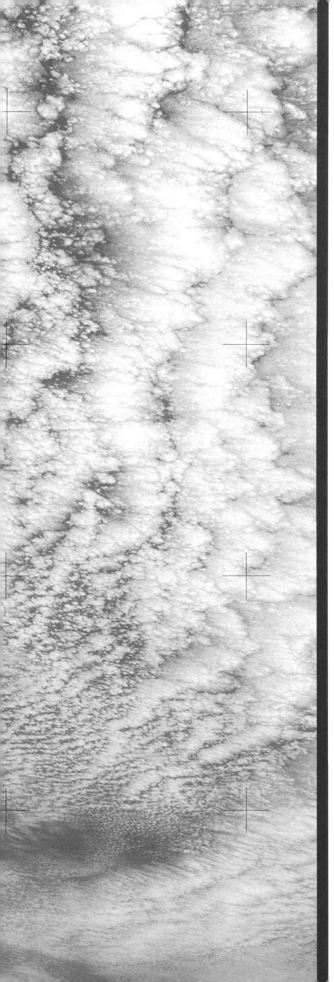

Chapter Nine

Skylab 1

The Saturn V had proved its mettle as a powerful and capable machine for the Apollo programme. As the lunar adventure began to wind down, the Saturn V had one more trick up its sleeve, but this time the iconic vehicle would have a central role – as a space station.

OPPOSITE Photographed by the last crew to visit, Skylab shows evidence of the damage that befell the station during its launch on 14 May 1973. Two parasols, one beneath the other, shield the workshop from solar heating and only a single solar cell wing extends from the side. A micrometeoroid shield came away during ascent and caused the eventual loss of the other wing. *(NASA)*

RIGHT Herman Potočnik (pseudonym Hermann Noordung, 1892–1929).

Dreams of a space station

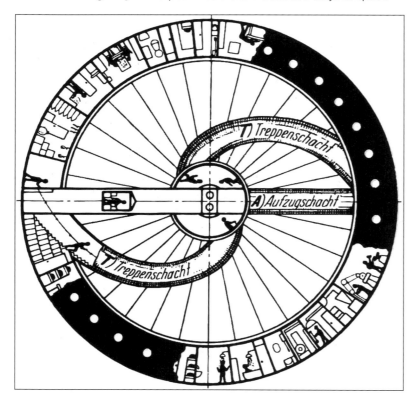

BELOW Noordung's concept of a rotating space station to provide artificial gravity.

Most people who had put serious thought into the idea of humans heading off-planet saw benefit in the construction of a habitable platform in Earth orbit. Such an outpost would allow extended stays in space and offer an opportunity for learning to live in this new environment. It could be a staging post for forays into deeper space, or it could be a manned base for the reconnaissance of the planet. Orbital stations were conceived in the early years of the twentieth century by notable pioneers of rocketry such as Konstantin Tsiolkovsky and Hermann Oberth.

Herman Potočnik, more commonly known by the surname Noordung, was an Austrian engineer who created advanced designs for a station that included the concept of rotation to make artificial gravity. His work was influential among the members of the German rocketry club, the VfR, that included Oberth and Wernher von Braun.

After being assimilated into the US, von Braun helped to write an influential series of articles for *Collier's* magazine in which he expressed Noordung's idea as a rotating 76m wheel that travelled from pole to pole at an altitude of 1,730km. This was before the discovery of what became known as the van Allen radiation belts, regions of trapped radiation that now force us to operate at lower altitudes. Von Braun envisaged that this station would give its occupants the ability to monitor the entire planet with powerful telescopes to inhibit an armed build-up anywhere in the world.

When the 'space race' between the Soviet Union and the United States began, the two participants kept the concept of a space station in mind as they worked to gain their first footholds in the new frontier. The advanced spacecraft that was then on NASA's drawing board, designated 'Apollo', would be part of a space infrastructure that could include voyages to the Moon, but a station in Earth orbit was seen as just as likely a destination. Such thoughts were, for the most part, pushed aside when President Kennedy laid down his lunar challenge. The Moon would occupy NASA for the coming decade.

NASA after Apollo

Even as the effort to put a man on the Moon was ramping up, some senior managers at NASA were aware of the institutional dangers posed by such a well-defined goal. Once it had been reached, how would they maintain the extraordinary capability that had been brought together under Apollo? It is a measure of the confidence of the times that even as vast resources were being spent on the lunar programme, many of NASA's centres were quietly studying the possibilities for an Earth-orbiting station.

It was clear that in order to save development costs, a station ought to use existing components if possible; maybe an enlarged Apollo spacecraft, or a separate station module accessed by an Apollo spacecraft. One suggestion from NASA's Langley Research Center, the *manned orbiting research laboratory* (MORL), envisaged a large, separate module serving as the space station but it would use existing Saturn IB or Titan boosters and the Gemini spacecraft that was then being developed in support of Apollo.

Air Force intentions

Concurrent with NASA's station-planning in the early to mid-1960s, the US Air Force was weighing up the military possibilities of an occupied platform on the assumption that, like reconnaissance aircraft of the time, any platform for looking down on Earth would require a crew to attend to the cameras and to develop photographic films. Powerful robotic reconnaissance satellites were still a long way off.

Attempts by the Air Force to assume NASA's role in space exploration were rebuffed, but they developed serious plans for a space glider called Dyna-soar and a military version of the Gemini spacecraft. When both of these ideas were shelved, the USAF concentrated on designing a space station of their own. This *manned orbiting laboratory* (MOL) would use elements from the Titan launch vehicle, essentially turning tanks into crew modules. A Gemini spacecraft would provide access. The development contract for MOL went to the Douglas Aircraft Company, giving them experience that would stand them in good stead.

In the event, MOL, too, was shelved as it became clear that the military's space needs could be met with ever more sophisticated unmanned satellites. The Air Force did manage a single launch in the MOL programme on 3 November 1966 when an unmanned Gemini spacecraft was lofted on top of a dummy

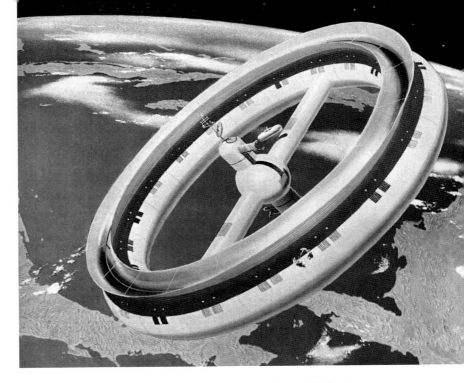

ABOVE Chesley Bonestell artwork from a 1952 edition of *Collier's* magazine illustrating Wernher von Braun's interpretation of Noordung's rotating space station. *(NASA-MSFC)*

BELOW An artist's rendition of the Manned Orbiting Laboratory, a USAF project that would use Titan launcher components and Gemini spacecraft. *(USAF)*

ABOVE George Mueller being sworn in by Hugh Dryden as the Associate Administrator for the Office of Manned Space Flight at NASA, September 1963. *(NASA)*

laboratory module that was nothing more than a spare Titan oxidiser tank.

In the meantime, in his role as the director of the Office for Manned Space Flight, George Mueller, the NASA manager who had masterminded the Saturn V's all-up testing regime, looked at NASA's possible options after the Moon landings were over. He assumed that the Apollo architecture would have uses beyond a lunar landing, and he was aware that it would take at least five years to get any project up and running. On 6 August 1965 he created the *Apollo applications programme* (AAP). It had an office in NASA headquarters, little definition of what it would do, and almost no money or support.

The orbital workshop

Prior to AAP, the Marshall Space Flight Center had also been casting around for a post-Apollo role in the knowledge that its background in very large civilian launch vehicles could not sustain it past Apollo. In the summer of 1965, they put forward the idea of converting a rocket stage into a space station after it had lifted itself into Earth orbit.

This wasn't a new idea. Every spacecraft enters space alongside the exhausted propulsive stage that got it there, and it had occurred to many engineers that the empty tanks of the largest rockets could be put to good use in orbit. Having put a lot of development effort into its S-IV stage, the Douglas Company made in-depth studies into its potential use as a space station, given that it had otherwise been superseded by the much improved S-IVB. Von Braun viewed these studies as the start of what became Skylab.

MSFC took the spent stage concept to the S-IVB itself, also a Douglas vehicle, and they hoped it would breathe more life into the Saturn IB, the vehicle for which it formed the upper stage. At the very least, this *orbital workshop* (as they had come to call it) would have a crew dock their Apollo spacecraft with the spent stage and then, having vented all the propellant, use the interior space to evaluate spacewalking techniques with the protection of the tank surrounding the astronaut. There were also ideas for pressurising the stage to make it into a habitable compartment.

The various possibilities being studied came to be divided into two broad types; a *wet workshop*, where a spent stage would be emptied of propellant in orbit prior to use, and a *dry workshop* that would be outfitted prior to launch and which would therefore never be used for propulsion.

Through late 1965 and on into 1966, NASA decided to look more closely at a wet workshop and asked the Manned Spacecraft Center (MSC, now JSC in Houston, Texas) to explore the idea with McDonnell, the company which had built the Mercury and Gemini spacecraft. Engineers were to study the requirements for a station that could be launched in 1968 on a Saturn IB. Their progress was rapid, but not without rancour because MSFC, which normally called the shots with rockets, and MSC, whose purview was spacecraft, struggled to divide responsibilities. On one hand, this was a rocket stage, but on the other, it would be transformed into a spacecraft.

In an attempt to add astronomical instruments that would be built into the shell of an Apollo lunar module, it became necessary to provide a means of docking multiple craft to the end of the empty second stage, an S-IVB. With the addition of an adapter to permit this, NASA was now looking at a space station that could be made much more flexible by having modules dock as necessary.

For the next two to three years, the project's progress was impeded by repeated threats to funding and arguments on how AAP should

RIGHT The tank assembly of S-IVB-212 during conversion into the orbital workshop of Skylab. *(NASA-MSFC)*

move forward. MSFC were convinced that the wet workshop could be made to work despite criticism from MSC that undermined the underlying assumptions. The Houston centre was sure that its modifications would improve the project's workability.

One idea in von Braun's mind was intended to follow a series of flights to the wet workshop. He proposed converting an S-IVB into a dry workshop that would be lofted in a fully fitted condition on top of a Saturn V. Towards the end of 1968 and into 1969, circumstances aligned to bring this radical idea to the fore.

With a change at the top, NASA's leadership became more disposed to a post-Apollo programme. Moreover, the spectacular success of Apollo 8 in December 1968 raised the public's enthusiasm for space endeavours and increased the likelihood that a Saturn V might become spare at the end of the Moon programme. The huge capability of this large rocket also permitted everything to be lofted at once. Multiple flights of the Saturn IB just to assemble the structure in orbit would be unnecessary. NASA's case for a substantial station further improved mid-1969 when the Pentagon cancelled the MOL project.

Birth of Skylab

On 22 July 1969, with Apollo 11 having fulfilled the central objective of the Apollo programme, NASA realigned its AAP effort towards a space station that would be converted from an S-IVB. This dry workshop would include an airlock, a docking adapter and a telescope facility. A scheme had been formulated whereby the telescope facility, rather than being a separate payload, would be mounted on a truss at the top of the stack, then rotated 90° to one side of the docking adapter once in orbit.

RIGHT The Skylab cluster being assembled. *(NASA-MSFC)*

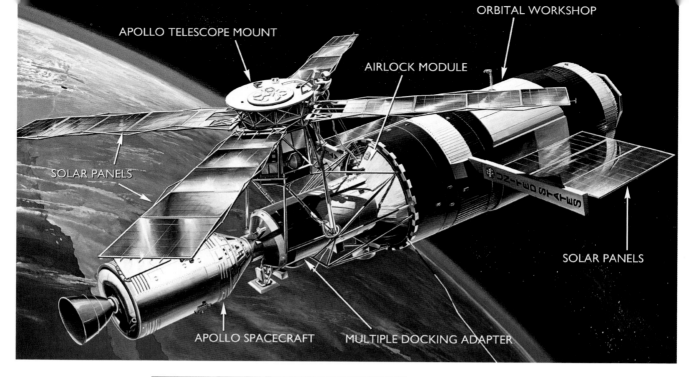

ABOVE Artist's impression of the Skylab workshop with an Apollo spacecraft attached. *(NASA-MSFC)*

RIGHT The distinctive latticework floors that separated the crew quarters from the forward and aft compartments. *(NASA-MSFC)*

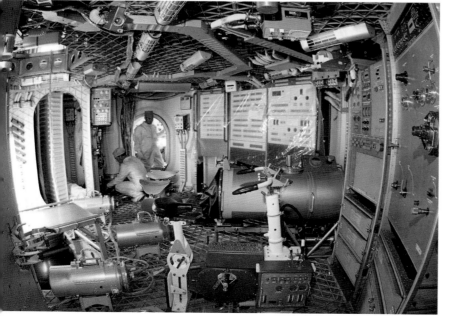

The four-year gestation of the orbital workshop had seen a remarkable development. An original modest concept for a spent rocket stage to serve as a useful working volume had become a $2.5 billion project to convert a large unused S-IVB into a fully fitted working space laboratory. With funding in place, NASA began to design and build flight hardware. They even gave the station a name: Skylab.

Description

Skylab consisted of four components; the *orbital workshop* (OWS), *airlock module* (AM), *multiple docking adapter* (MDA) and *Apollo telescope mount* (ATM). All four were launched as a cluster by a two-stage Saturn V. The cluster was visited on three occasions by Apollo spacecraft for stays that had increasing durations of 28, 56 and 84 days.

The workshop was a reconfigured S-IVB stage whose hydrogen tank became a living and working space for three astronauts. Towards the aft end of the tank, two bulkheads split the space into three; a forward compartment and dome, crew quarters, and aft compartment. The crew quarters were further

LEFT The experiment area of the crew quarters, seen during preparation. *(NASA-MSFC)*

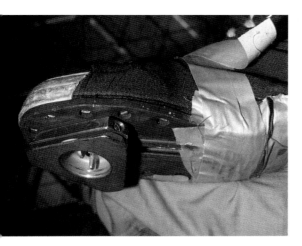

ABOVE **One of the shoes worn by the crew, showing the cleats that engaged with the lattice floor to give an anchor point.** *(NASA-MSFC)*

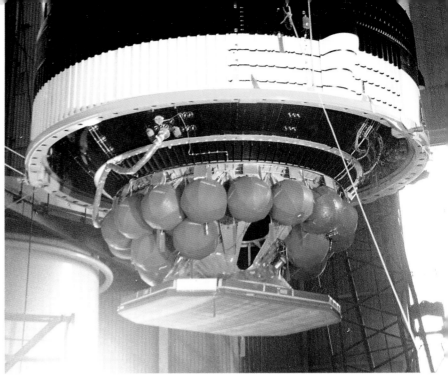

ABOVE **The aft end of the workshop had 23 nitrogen-filled spheres, seen here preflight with protective wrapping. 22 provided propellant for attitude control and the remaining sphere supplied pneumatic pressure.** *(NASA-MSFC)*

subdivided into compartments for sleep, waste management, experiments, and a wardroom. The floor and ceiling structures were faced with an aluminium grid featuring a triangular pattern to engage with cleats on the crewmen's shoes as a means of anchoring them in place.

The LOX tank was converted into a storage space for both solid and liquid waste. Filter screens were used to compartmentalise the tank to keep solid and liquid separate. Because the tank was unpressurised, liquid waste would evaporate or sublime to vapour and escape to space via non-propulsive vents.

Further aft, the thrust structure became a carrier for 23 spheres that stored high pressure nitrogen that provided pneumatic pressure where needed and a source gas for the *thruster attitude control subsystem* (TACS), which was one of Skylab's two means of attitude control. A micrometeoroid shield encased the thrust structure. There being no J-2 engine present, a large, flat radiator was installed there to dissipate waste heat.

The station's cylindrical wall was wrapped in a micrometeoroid shield of 0.6mm aluminium sheet. This would be held close against the workshop for launch and be deployed to a standoff distance of 13cm in orbit. Two airlocks were provided to allow science experiments to be sent outside through 21cm apertures in Skylab's hull.

Electrical power would be provided chiefly by two huge solar panels that were folded against the side of the station for launch. These would deploy in orbit to provide about 10.5kW. Up to

BELOW **One of the four solar panels that would be extended from the Apollo telescope mount.** *(NASA-MSFC)*

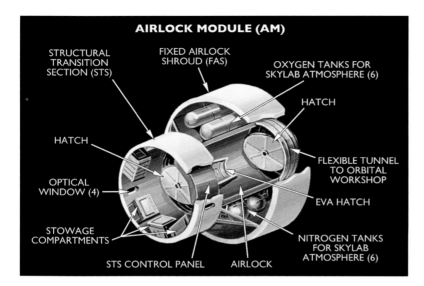

LEFT Diagram of the airlock module. (NASA/Woods)

CENTRE The airlock module at McDonnell Douglas. The oddly shaped opening is for the Gemini EVA hatch that was repurposed for use on Skylab. Gemini had been a McDonnell spacecraft. (NASA-MSFC)

about 5kW of additional power was to come from a set of panels deployed on the ATM. A portion of the electricity generated from the Sun would power the station while on Earth's daytime side and the rest would be stored in batteries to provide power during night-time passes.

Forward of the workshop was the airlock module, a 5.3m component comprising a tunnel and a structural collar that led to the MDA. It carried a range of equipment, primarily for atmosphere control and electrical distribution. A truss connected the airlock to a cylindrical shroud that was essentially an extension of the stage's outline forward of the instrument unit. This shroud would come in useful for Skylab's rehabilitation during its first occupation.

As its name suggests, the airlock module could be depressurised to allow suited astronauts to exit the station. The exit hatch was a surplus Gemini spacecraft hatch complete with a window. Rather than design from scratch, engineers simply borrowed this outward-opening design that had proved itself on a previous programme.

The MDA was a 5.2m cylindrical structure, 3m in diameter. It carried two docking ports, one axial for normal use and the other to the side in case a fault disabled the docked Apollo spacecraft and it became necessary to launch a rescue mission.

The interior of the MDA was a control centre for much of Skylab's science experiments including the telescopes in the ATM, a package for Earth resource science, and others that studied metal and material processing in microgravity.

Derived loosely from the octagonal form of the Apollo lunar module descent stage,

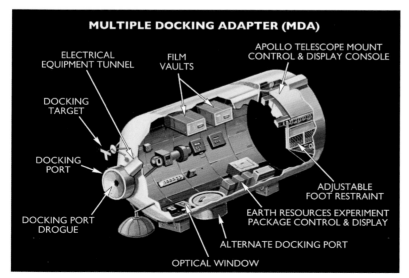

LEFT Diagram of the multiple docking adapter. (NASA/Woods)

RIGHT The multiple docking adapter being given weight and balance checks at MSFC. *(NASA-MSFC)*

the ATM held a suite of solar telescopes and equipment for stabilising the station and the telescopes. For launch it would be carried by a truss in front of the MDA. In orbit, cables would be reeled to rotate the entire structure through 90° and then it would deploy an X-shaped array of four solar panels to augment the station's power.

The ATM housed three control moment gyros (CMG) as Skylab's primary attitude control system. At the core of a CMG was a heavy disk (65.5kg) that spun at about 9,000rpm. By virtue of a gyro's property of resisting rotation of its axis, it provided a stable base against which the station could be rotated without using up irreplaceable thruster gas. Two were sufficient to control the station and a third provided backup. The telescopes themselves had an additional pointing system to achieve the very accurate stability they required.

RIGHT Diagram of the Apollo telescope mount. *(NASA/Woods)*

BELOW The octagonal form of a lunar module descent stage is evident in the Apollo telescope mount, here being moved to a clean room at MSFC. *(NASA-MSFC)*

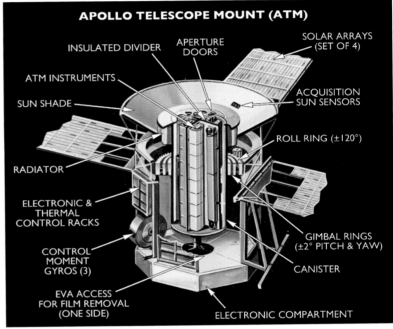

LEFT The objective end of the eight solar instruments on the Apollo telescope mount. *(NASA-MSFC)*

ABOVE Launch of Skylab from Pad 39A on 14 May 1973. *(NASA)*

Skylab 1's near-disaster

On 14 May 1973, the orbital workshop was launched from Pad 39A at the Kennedy Space Center towards a 50° orbit. That is, as it coasted around the planet, the plane of the orbit would form a 50° angle with respect to the equator and its ground track's most northerly and southerly points would be at latitudes 50°N and 50°S. This orbit was chosen to take the station over more populated land than was usual for an Apollo mission.

As detailed on page 147, Skylab's ascent didn't quite go as planned. However, the S-IC and S-II stages of the Saturn V, flying their final mission, did a fine job.

In orbit, the ATM was rotated to the side of the docking adapter and its solar panels were deployed to form a distinctive 'X'. However, flight controllers soon received indications that the workshop's two wing-like solar panels were not deploying properly. Then as engineers inspected the telemetry from the launch vehicle, they gained evidence that something had gone seriously awry.

A design flaw had caused Skylab's micrometeoroid shield to rip away in the first minute of flight. This calamity deprived the station of protection, not only from small, high-speed projectiles (not considered a huge problem) but also from the Sun's direct heat. Worse, one of the two main solar panels had been ripped clean off by the blast from the S-II's retrorockets. The other panel was unable to deploy.

The launch of the first crew, planned for 24 hours later from Pad 39B, was successively postponed while teams of engineers went into overdrive to work on the problems. If they orientated the station end-on to the Sun in order to reduce heating, they got no electrical power from the solar panels on the ATM. They found a compromise attitude which still left the interior of the workshop at 42°C. Meanwhile, the airlock, which included water cooling systems for spacesuits, became so cold that it threatened to freeze, possibly rupturing components.

Just when the engineers needed accurate information on the workshop's attitude, they discovered that the gyroscopes that measured its rotation were overheating and giving erratic measurements. Alternative indicators of attitude were pressed into service; the relative temperatures on opposite sides of the station and the electrical output from the ATM solar panels.

BELOW The partially deployed arm of the remaining solar panel on the orbital workshop. The ragged edge of the torn micrometeoroid shield is evident. *(NASA)*

ABOVE In order to launch a crew to Skylab, it was necessary to mount the Saturn IB on a 'milkstool' to mate it with the facilities designed for the Saturn V. This is Skylab 4 leaving Pad 39B on 16 November 1973 *(NASA)*

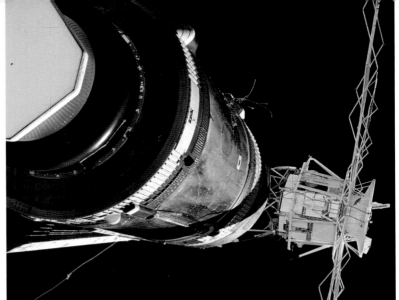

TOP Skylab 2's view of the damaged workshop. Torn wires reveal where a solar panel arm was ripped away. *(NASA)*

ABOVE The discoloured sunward side of the workshop. The root of the missing arm is to the right. At the bottom is the square aperture of the science airlock from where a parasol would be deployed. *(NASA)*

The rescue of Skylab

The Skylab 2 mission with the first crew was launched after a ten-day postponement during which teams at MSFC, MSC and various contractors and institutions had cast around for a solution to the loss of the shield. Two schemes had come to the fore. In both, a 'parasol' 6.7m × 7.3m would be deployed over the sunward side of the workshop.

One proposal was to support the parasol using two extensible poles that would be attached to the ATM during a spacewalk. This would be a backup plan. The primary proposal was to deploy the parasol through the science airlock on the sunward side of the station which included a facility to operate equipment externally and therefore could be used to unfurl the parasol

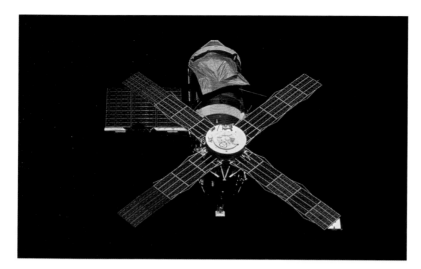

LEFT Skylab as seen on departure of the first crew. The parasol covers the sunward side of the station to keep temperatures down. *(NASA)*

once it had passed outside without requiring the astronauts to undertake a spacewalk.

As Skylab 2 closed in on the station, Pete Conrad, the spacecraft's commander, confirmed that one of the two main solar panels was completely missing and the second had partially deployed. "There's a bulge of meteoroid shield underneath it in the middle, and it looks to be holding it down." An initial attempt by Paul Weitz to free the panel while standing in the hatch of the Apollo spacecraft failed.

After some docking difficulties Conrad, Weitz and Joe Kerwin entered the station and began to deal with its issues. The primary task was to deploy the parasol through the scientific airlock. This was achieved without difficulty, and over the next four days the workshop's internal temperature fell to tolerable levels.

The electrical power supply was critical because the ATM couldn't supply sufficient energy for the station's needs. Conrad and Kerwin had to go outside to release the remaining solar panel from the side of the workshop.

The scheme that had been worked out on Earth required Kerwin to attach cutters to a long pole and then use this to cut the debris that was impeding the panel's deployment. Next Conrad used the pole to reach the panel's

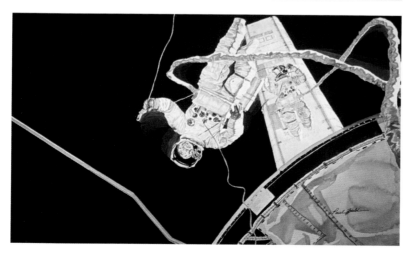

CENTRE This painting by artist Paul Fjeld was produced after Skylab 2's launch but prior to the repair EVA. It shows Pete Conrad using a tether to attempt to raise the stuck solar panel arm. Its accuracy was confirmed by astronaut Rusty Schweickart who had perfected the technique in a water tank. *(NASA)*

LEFT Upon the return of the Skylab 2 crew, artist Paul Fjeld used careful research to produce this painting. Conrad and Kerwin (foreground) are shown cartwheeling away a few seconds after the stuck arm had come free. *(NASA)*

LEFT The Skylab 3 crew deployed the backup parasol on top of the original, further improving conditions within the workshop. (NASA)

arm, where he hooked up a cable to one of its vent covers. The other end of the cable was hooked to the edge of the airlock's shroud. Finally, both crewmen lifted the centre of the cable away from the hull, thereby exerting sufficient force to overcome a damping mechanism that had frozen.

Conrad related, "I was facing away from it, heaving with all my might and Joe was also heaving with all his might when it let go and both of us took off.... By the time we got settled down and looked at it, those panels were out as far as they were going to go at the time."

Later, Conrad said that he considered his part in rescuing Skylab to be his greatest moment in space, even better than walking on the Moon. "The Moon did not tax me, nor give me as much satisfaction as Skylab." The crew completed their planned stay of 28 days and left the station in a fit state for subsequent crews.

Skylab 3 deployed the backup two-pole parasol over the original sunshade, further improving conditions inside the station. Difficulties with the early part of the mission made the crew resolve to more than catch up during their remaining time. They proved to be machine-like in their quest to be over-achievers by the time their 56-day stay was complete.

Skylab's final occupation lasted 84 days, but this crew found it hard to live up to the pace set by their predecessors. The flight planners wanted to squeeze in as much science as possible, but they had difficulty

LEFT The Gemini programme had failed to test an astronaut manoeuvring unit. Here, Skylab 3 commander, Alan Bean, tests a similar device within the safe confines of the workshop. (NASA)

BELOW Scientist-astronaut Joe Kerwin carries out an oral examination of Skylab 2 commander Pete Conrad. (NASA)

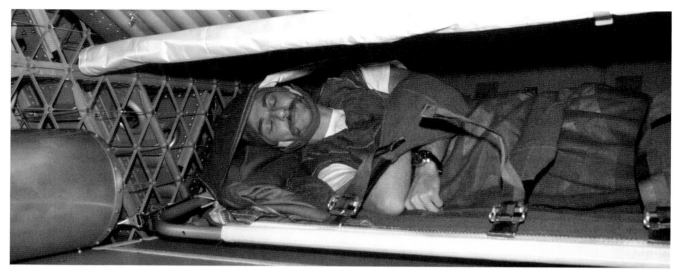

ABOVE Skylab 3 scientist-astronaut Owen Garriot restrained in his sleeping bunk. *(NASA)*

BELOW A view along the length of the orbital workshop during the Skylab 4 mission. At the far end are astronauts Ed Gibson and Gerald Carr. *(NASA)*

OPPOSITE The Skylab cluster orbiting over Earth, as photographed by the Skylab 2 crew as they departed. *(NASA)*

properly accounting for the time taken to complete tasks in space in comparison to doing so on the ground. The crew had come to realise that completing a task properly was more important than completing it on time. As soon as an astronaut fell behind on one task he was not able to recover the schedule and the remainder of his day became one long frustration. To make their point, the astronauts

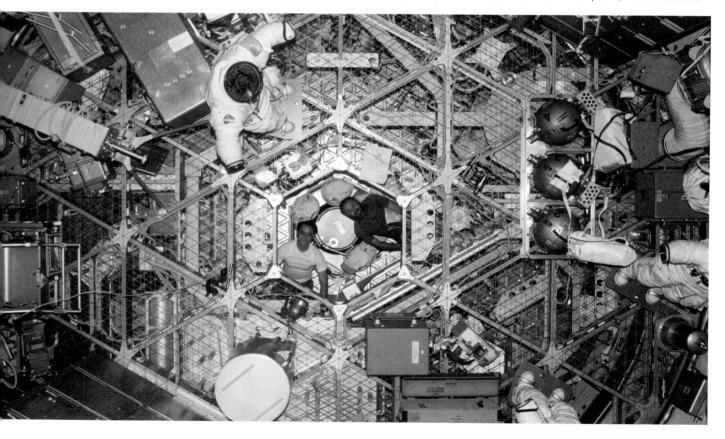

went on strike until the planners relaxed the pace. Ironically, with this new regime in play, the crew were able to approach the productivity of their predecessors flying much shorter missions.

Skylab's demise

After the Skylab 4 crew departed on 8 February 1974, the station was left in a mothballed state, controlled from Earth. Without propulsion and still within the upper traces of the atmosphere, its orbit was slowly decaying. NASA desired to reuse it in the future and hoped that the Space Shuttle, then in development, would be ready in time to dock with Skylab and boost it into a higher orbit from where it could become a destination for crews.

Increased solar activity during the 1970s caused Earth's atmosphere to expand slightly, and the station's orbit began to decay faster than anticipated. On 11 July 1979 it re-entered over Western Australia. It surprised engineers by taking much longer than anticipated to break up, scattering pieces across that continent.

Though it met an ignominious end, the Saturn S-IVB that became Skylab provided an important lesson in prolonged human habitation of space, one that would be carried forward to its eventual expression in the huge International Space Station. Certainly using up an S-IVB and a Saturn V launch vehicle in this way was preferable to turning them into lawn ornaments.

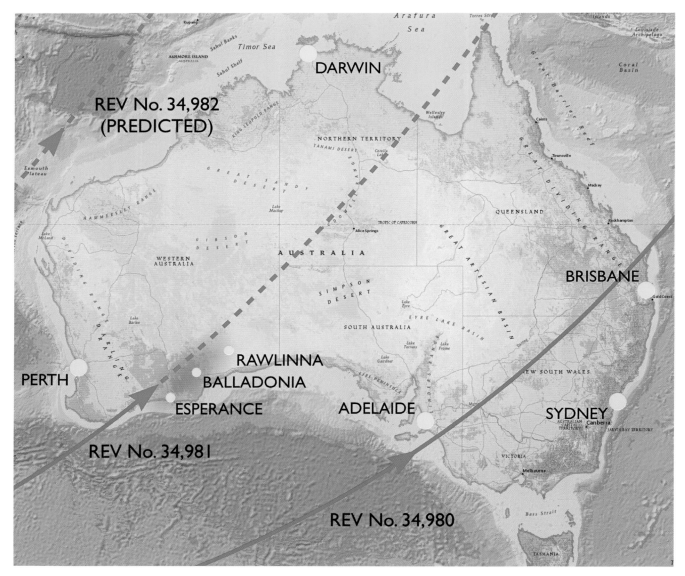

BELOW During its 34,981st orbit, Skylab re-entered the atmosphere, scattering debris across a wide area of Western Australia. *(Woods/Map from National Geographic's MapMaker Interactive)*